INTRODUCTION TO GRID AND CLOUD COMPUTING

Dr. R. Deepalakshmi
Professor
Dr. P. Alli
Professor and Head of the Department
Ms. M. Shyni Beaulah
Assistant Professor
Department of CSE
Velammal College of Engineering and Technology, Madurai

SHANLAX PUBLICATIONS
MADURAI, Tamil Nadu, INDIA

INTRODUCTION TO GRID AND CLOUD COMPUTING

First Edition: 2017

ISBN: 978-93-86537-38-6

Price: ₹ 410/-

Publisher
SHANLAX PUBLICATIONS
61, 66 T.P.K. Main Road
Vasantha Nagar
MADURAI – 625003
Tamil Nadu, INDIA

Ph: 0452-4208765,
Mobile: 7639303383
email: publisher@shanlaxpublications.com
web: www.shanlaxpublications.com

Anna University Regulation 2013
Syllabus CS 6703 Introduction to Grid and Cloud Computing

TOPICS TO BE COVERED

UNIT I INTRODUCTION **9**

Evolution of Distributed computing: Scalable computing over the Internet – Technologies for network based systems – clusters of cooperative computers - Grid computing Infrastructures – cloud computing - service oriented architecture – Introduction to Grid Architecture and standards – Elements of Grid – Overview of Grid Architecture.

UNIT II GRID SERVICES **9**

Introduction to Open Grid Services Architecture (OGSA) – Motivation – Functionality Requirements – Practical & Detailed view of OGSA/OGSI – Data intensive grid service models – OGSA services.

UNIT III VIRTUALIZATION **9**

Cloud deployment models: public, private, hybrid, community – Categories of cloud computing: Everything as a service: Infrastructure, platform, software - Pros and Cons of cloud computing – Implementation levels of virtualization – virtualization structure – virtualization of CPU, Memory and I/O devices – virtual clusters and Resource Management – Virtualization for data center automation.

UNIT IV PROGRAMMING MODEL **9**

Open source grid middleware packages – Globus Toolkit (GT4) Architecture , Configuration – Usage of Globus – Main components and Programming model - Introduction to Hadoop Framework - Mapreduce, Input splitting, map and reduce functions, specifying input and output parameters, configuring and running a job – Design of Hadoop file system, HDFS concepts, command line and java interface, dataflow of File read & File write.

UNIT V SECURITY **9**

Trust models for Grid security environment – Authentication and Authorization methods – Grid security infrastructure – Cloud Infrastructure security: network, host and application level – aspects of data security, provider data and its security, Identity and access management architecture, IAM practices in the cloud, SaaS, PaaS, IaaS availability in the cloud, Key privacy issues in the cloud.

Total : 45

COURTESY

1. Kai Hwang, Geoffery C. Fox and Jack J. Dongarra, "Distributed and Cloud Computing: Clusters, Grids, Clouds and the Future of Internet", First Edition, Morgan Kaufman Publisher, an Imprint of Elsevier, 2012
2. Jason Venner, "Pro Hadoop- Build Scalable, Distributed Applications in the Cloud", A Press, 2009
3. Tom White, "Hadoop The Definitive Guide", First Edition. O"Reilly, 2009.
4. Bart Jacob (Editor), "Introduction to Grid Computing", IBM Red Books, Vervante, 2005
5. Ian Foster, Carl Kesselman, "The Grid: Blueprint for a New Computing Infrastructure", 2nd Edition, Morgan Kaufmann.
6. Frederic Magoules and Jie Pan, "Introduction to Grid Computing" CRC Press, 2009.
7. Daniel Minoli, "A Networking Approach to Grid Computing", John Wiley Publication, 2005.
8. Barry Wilkinson, "Grid Computing: Techniques and Applications", Chapman and Hall, CRC, Taylor and Francis Group, 2010.

Contents

1. INTRODUCTION

1.1 Evolution of Distributed computing

Computing since its inception has undergone many changes, from large computers that perform tasks allowed to a limited and somewhat exclusive use very select organizations, to present either personal computers or laptops that have the same or even greater capabilities than the first and are increasingly introduced into the daily life of a person.

The biggest changes are mainly attributed to two causes, which occurred from the seventies:

1. The development of microprocessors, which led to a reduction in size and cost of computers and greatly increase the capabilities of themselves and their access to more people.
2. The development of local area networks and communication for connecting computers can transfer data at high speed.

1.1.1 Development of Distributed Systems

Definition:

"Systems whose hardware and software components, which are on networked computers, communicate and coordinate their actions by passing messages, to achieve a goal. The connection is made using a predetermined protocol scheme for a client-server".

Features:

- **Concurrency** - This feature allows distributed systems resources available in the network can be used simultaneously by users and / or agents that interact in the network.
- **Lack of global clock** - The coordination for the transfer of messages between the various components to perform a task, do not have a general timing, is more evenly distributed to the components.
- **Independent failures of components** - Each system component may fail independently with which others can continue to execute their actions. This allows the achievement of the tasks more effectively, because the whole system is still working.

Evolution:

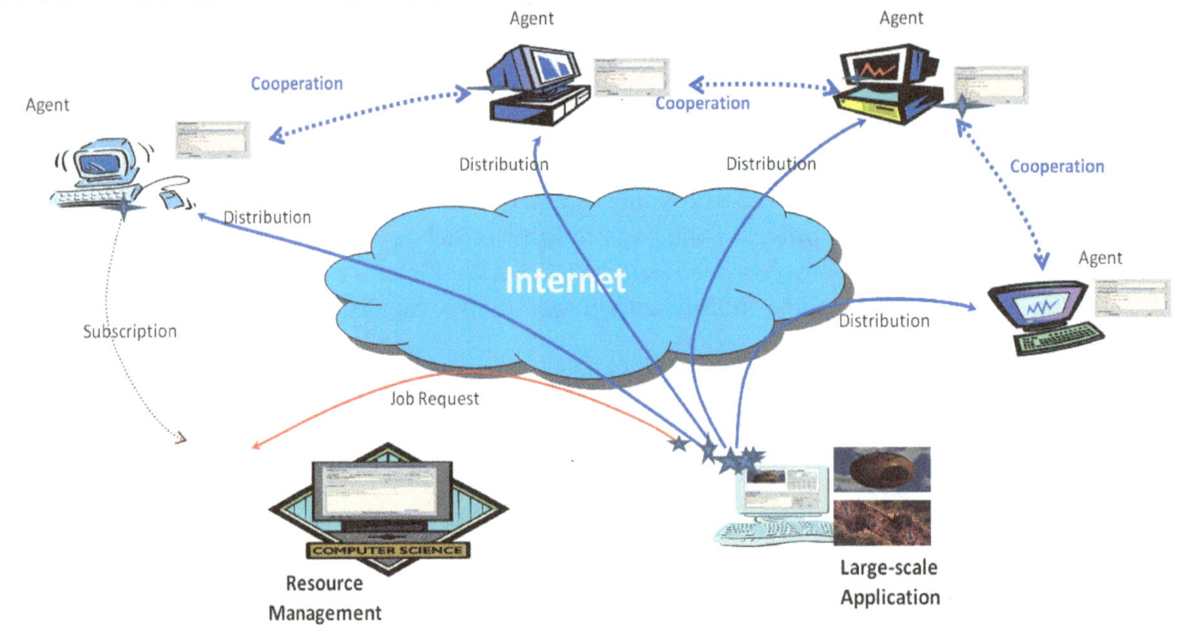

Figure 1.1 Distributed Systems

Central Processing (Host) - One of the first models of interconnected computers, call center, where all the processing of the organization are carried out on a single computer, usually a mainframe, and personal computer users employed simple. The problems with this model are 1.when the increased processing load had to change the hardware of the mainframe, which is more expensive to add more client computers or servers to increase capacity.2.The other problem that arose is the modern graphical user interfaces, which could lead to a large increase in traffic on the media and therefore could collapse.

Server Group - Another model that came to compete with the former, also somewhat centralized, are a group of computers acting as servers, normally files or print, unintelligent for a number of minicomputers that are connected to a processing local area network. The problems with this model is Could be generated saturation media between servers and minicomputers unintelligent, for example, when files are requested grades for several customers at once, could greatly reduce the speed of transmission.

Client Server Computing - This model, which prevails at present, can decentralize the processing and resources, above all, of each of the services and the Display of Graphical User Interface. This means that some servers are dedicated only to a particular application and therefore run efficiently.

Definition:

System where the client is a machine requesting a particular service and server is called the machine that is provided. The services include:

* Implementation of a program.
* Access to specific bank information.
* Access to a hardware device.

It is an essential element, the presence of a physical means of communication between machines, and will depend on the nature of this medium the viability of the system.

1.2 Scalable Computing Over the Internet

Over the past 60 years, computing technology has undergone a series of platform and environment changes. We assess evolutionary changes in machine architecture, operating system platform, network connectivity, and application workload. Instead of using a centralized computer to solve computational problems, a parallel and distributed computing system uses multiple computers to solve large-scale problems over the Internet. Thus, distributed computing becomes data-intensive and network-centric. This section identifies the applications of modern computer systems that practice parallel and distributed computing. These large-scale Internet applications have significantly enhanced the quality of life and information services in society today.

1.2.1 The Age of Internet Computing

Billions of people use the Internet every day. As a result, supercomputer sites and large data centers must provide high-performance computing services to huge numbers of Internet users concurrently. The emergence of computing clouds instead demands high-throughput computing systems built with parallel and distributed computing technologies. We have to upgrade data centers using fast servers, storage systems, and high-bandwidth networks. The purpose is to advance network-based computing and web services with the emerging new technologies.

1.2.1.1 The Platform Evolution

Computer technology has gone through five generations of development,

- 1950 to 1970- handfuls of mainframes were built to satisfy the demands of large businesses and government organizations. Ex. IBM 360 and CDC 6400
- 1960 to 1980- lower cost minicomputers became popular among small businesses and on college campuses. Ex. DEC PDP 11 and VAX Series
- 1970 to 1990- personal computers built with VLSI microprocessors.
- 1980 to 2000- massive numbers of portable computers and pervasive devices appeared in both wired and wireless applications.
- Since 1990, the use of both HPC and HTC systems hidden in clusters, grids, or Internet clouds has proliferated. These systems are employed by both consumers and high-end web-scale computing and information services.

The general computing trend is to leverage shared web resources and massive amounts of data over the Internet. Figure 1.1 illustrates the evolution of High-Performance Computing (HPC) and High-Throughput Computing (HTC) systems. On the HPC side, supercomputers (massively parallel processors or MPPs) are gradually replaced by clusters of cooperative computers out of a desire to share computing resources. The cluster is often a collection of homogeneous compute nodes that are physically connected in close range to one another.

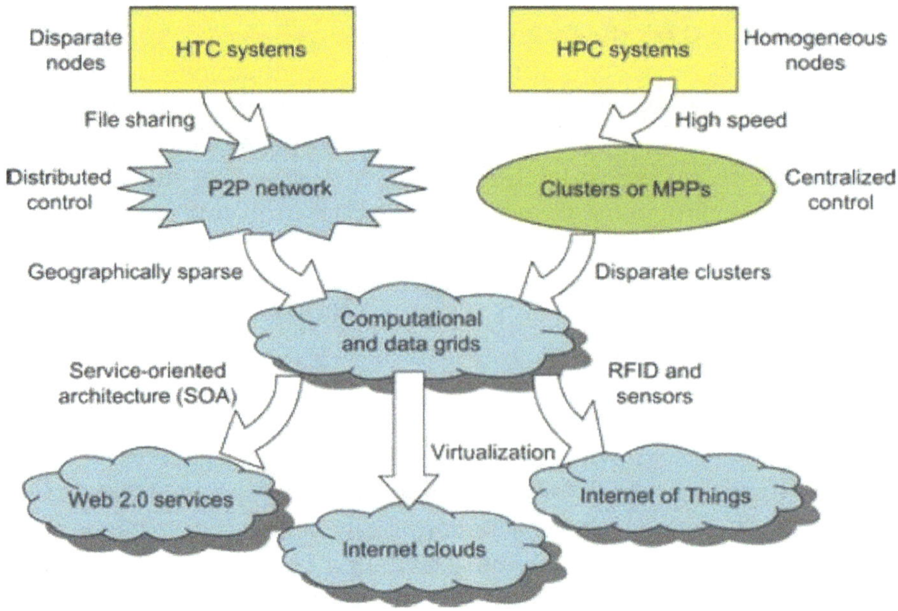

Figure 1.2 Evolutionary trend toward parallel, distributed, and cloud computing

1.2.1.2 High-Performance Computing

High-performance computing

- Is the use of parallel processing for running advanced application programs efficiently, reliably and quickly.
- The term applies especially to systems that function above a teraflop or 10^{12} floating-point operations per second.
- Technically a supercomputer is a HPC system that performs at or near the currently highest operational rate for computers.
- Supercomputers work at more than a petaflop or 10^{15} floating-point operations per second. This improvement was driven mainly by the demands from scientific, engineering, and manufacturing communities.

1.2.1.3 High-Throughput Computing

- The development of market-oriented high-end computing systems is undergoing a strategic change from an HPC paradigm to an HTC paradigm.
- HTC paradigm pays more attention to high-flux computing. The main application for high-flux computing is in Internet searches and web services by millions or more users simultaneously.
- The performance goal thus shifts to measure high throughput or the number of tasks completed per unit of time.
- HTC technology needs to not only improve in terms of batch processing speed, but also address the acute problems of cost, energy savings, security, and reliability at many data and enterprise computing centers.

1.2.1.4 Three New Computing Paradigms

- Virtualization
- Cloud computing
- Internet of Things

Advances in virtualization make it possible to see the growth of Internet clouds as a new computing paradigm. The maturity of radio-frequency identification (RFID), Global Positioning System (GPS), and sensor technologies has triggered the development of the Internet of Things (IoT).

1.2.1.5 Computing Paradigm Distinctions

The following list defines the terms Centralized computing, Parallel computing, Distributed computing and Cloud computing more clearly

Centralized computing:

- In this paradigm all computer resources are centralized in one physical system.
- All resources (processors, memory, and storage) are fully shared and tightly coupled within one integrated OS.
- Many data centers and supercomputers are centralized systems.

Parallel computing

- In this all processors are either tightly coupled with centralized shared memory or loosely coupled with distributed memory.
- Interprocessor communication is accomplished through shared memory or via message passing.
- This is referred as parallel processing. A computer system capable of parallel computing is commonly known as a parallel computer.
- Programs running in a parallel computer are called parallel programs. The process of writing parallel programs is often referred to as parallel programming.

Distributed computing

- A distributed system consists of multiple autonomous computers, each having its own private memory, communicating through a computer network.
- Information exchange is accomplished through message passing.
- A computer program that runs in a distributed system is known as a distributed program.
- The process of writing distributed programs is referred to as distributed programming.

Cloud computing

- An Internet cloud of resources can be either a centralized or a distributed computing system.
- The cloud applies parallel or distributed computing, or both.
- Clouds can be built with physical or virtualized resources over large data centers that are centralized or distributed.
- Cloud computing to be a form of utility computing or service computing.

Other Computing Paradigm

- **Concurrent computing or concurrent programming** typically refer to the union of parallel computing and distributing computing, although biased practitioners may interpret them differently.
- **Ubiquitous computing** refers to computing with pervasive devices at any place and time using wired or wireless communication. The Internet of Things (IoT) is a networked connection of everyday objects including computers, sensors, humans, etc. The IoT is supported by Internet clouds to achieve ubiquitous computing with any object at any place and time.
- **Internet computing** is even broader and covers all computing paradigms over the Internet.

1.2.1.6 Distributed System Families

Since the mid-1990s, technologies for building P2P networks and networks of clusters establish wide area computing infrastructures, known as computational grids or data grids. Internet clouds are the result of moving desktop computing to service-oriented computing using server clusters and huge databases at data centers. Grids and clouds are disparity systems that place great emphasis on resource sharing in hardware, software, and data sets. Massively distributed systems are intended to exploit a high degree of parallelism or concurrency among many machines.

In October 2010, the highest performing cluster machine was built in China with 86016 CPU processor cores and 3,211,264 GPU cores in a Tianhe-1Asystem. The largest computational grid connects up to hundreds of server clusters. A typical P2P network may involve millions of client machines working simultaneously. Experimental cloud computing clusters have been built with thousands of processing nodes.

In the future, both HPC and HTC systems will demand multicore or many-core processors that can handle large numbers of computing threads per core. Both HPC and HTC systems emphasize parallelism and distributed computing. Future HPC and HTC systems must be able to satisfy this huge demand in computing power in terms of throughput, efficiency, scalability, and reliability. The system efficiency is decided by speed, programming, and energy factors (i.e., throughput per watt of energy consumed). Meeting these goals requires yielding the following design objectives:

- **Efficiency** measures the utilization rate of resources in an execution model by exploiting massive parallelism in HPC. For HTC, efficiency is more closely related to job throughput, data access, storage, and power efficiency.
- **Dependability** measures the reliability and self-management from the chip to the system and application levels. The purpose is to provide high-throughput service with Quality of Service (QoS) assurance, even under failure conditions.
- **Adaptation** in the programming model measures the ability to support billions of job requests over massive data sets and virtualized cloud resources under various workload and service models.
- **Flexibility** in application deployment measures the ability of distributed systems to run well in both HPC (science and engineering) and HTC (business) applications.

1.2.2 Scalable Computing Trends and New Paradigms

Several predictable trends in technology are known to drive computing applications. In fact, designers and programmers want to predict the technological capabilities of future systems.

1.2.2.1 Degrees of Parallelism

- **Bit-Level Parallelism** (BLP) converts bit-serial processing to word level processing gradually.
- **Instruction-Level Parallelism** (ILP) the processor executes multiple instructions simultaneously rather than only one instruction at a time.
    - ✓ ILP practiced through pipelining, superscalar computing, VLIW (very long instruction word) architectures, and multithreading.
    - ✓ ILP requires branch prediction, dynamic scheduling, speculation, and compiler support to work efficiently.
- **Data-Level Parallelism** (DLP) was made popular through SIMD (single instruction, multiple data) and vector machines using vector or array types of instructions. DLP requires even more hardware support and compiler assistance to work properly.
- **Task-Level Parallelism** (TLP) was made popular when multicore processors and chip multiprocessors (CMPs) are introduced.
- **Job-Level parallelism** (JLP) as we move from parallel processing to distributed processing, we will see an increase in computing granularity to **job-level parallelism** (JLP). It is fair to say that coarse-grain parallelism is built on top of fine-grain parallelism.

A modern processor explores all of the aforementioned parallelism types. In fact, BLP, ILP, and DLP are well supported by advances in hardware and compilers. However, TLP is far from being very successful due to difficulty in programming and compilation of code for efficient execution on multicore CMPs.

1.2.2.2 Innovative Applications

Both HPC and HTC systems desire transparency in data access, resource allocation, process location, concurrency in execution, job replication, and failure recovery to both users and system management. Table 1.1 highlights a few key applications that have driven the development of parallel and distributed systems over the years. All applications demand computing economics, web-scale data collection, system reliability, and scalable performance. For example, distributed transaction processing is often practiced in the banking and finance industry. Transactions represent 90 percent of the existing market for reliable banking systems. Users must deal with multiple database servers in distributed transactions. Maintaining the consistency of replicated transaction records is crucial in real-time banking services. Other complications include lack of software support, network saturation, and security threats in these applications.

Table 1.1 Applications of High-Performance and High-Throughput Systems

Domain	Specific Applications
Science and Engineering	❖ Scientific simulation, genomic analysis etc ❖ Earthquake prediction, global warming, weather forecasting etc
Business, Education, Service Industry and Health	❖ Telecommunication, content delivery, e-commerce etc ❖ Banking, stock exchange, transaction processing etc ❖ Air care traffic control, electric power grids, Distance education etc ❖ Health care, hospital automation, telemedicine etc
Internet web services and Government applications	❖ Internet search, data centers, decision making system etc ❖ Traffic monitoring, worm containment, cyber security etc ❖ Digital government, online tax return processing, social networking etc
Mission Critical applications	❖ Military command and control, intelligent systems, crisis management

1.2.2.3 The Trend toward Utility Computing

Figure 1.2 identifies major computing paradigms to facilitate the study of distributed systems and their applications. These paradigms share some common characteristics.

- First, they are all ubiquitous in daily life. Reliability and scalability are two major design objectives in these computing models.
- Second, they are aimed at autonomic operations that can be self-organized to support dynamic discovery.
- Finally, these paradigms are composable with QoS and SLAs (service-level agreements).

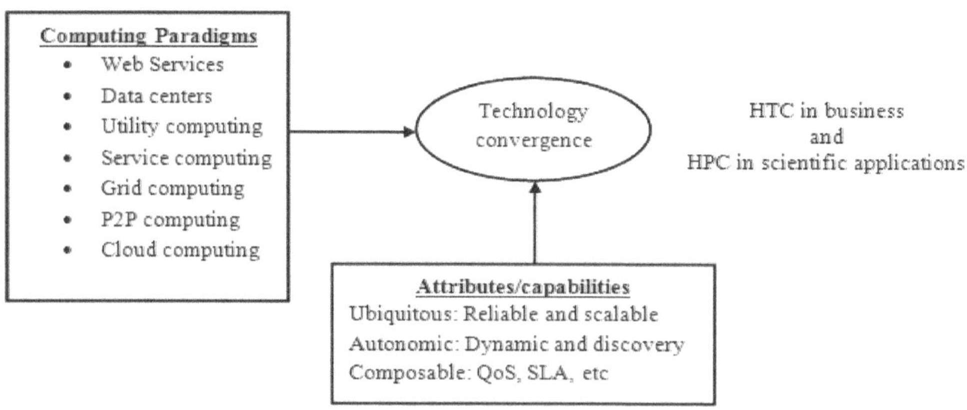

Figure 1.3 The vision of computer utilities in modern distributed computing systems

Utility computing focuses on a business model in which customers receive computing resources from a paid service provider. All grid/cloud platforms are regarded as utility service providers. However, cloud computing offers a broader concept than utility computing. Distributed cloud applications run on any available servers in some edge networks. Major technological challenges include all aspects of computer science and engineering. For example, users demand new network efficient processors, scalable memory and storage schemes, distributed OS, middleware for machine virtualization, new programming models, effective resource management, and application program development. These hardware and software supports are necessary to build distributed systems that explore massive parallelism at all processing levels.

1.2.2.4 The Hype Cycle of New Technologies

Any new and emerging computing and information technology may go through a hype cycle, as illustrated in Figure 1.3. This cycle shows the expectations for the technology at five different stages.

- The expectations rise sharply from the trigger period to a high peak of inflated expectations.
- Through a short period of disillusionment, the expectation may drop to a valley and then increase steadily over a long enlightenment period to a plateau of productivity.
- The number of years for an emerging technology to reach a certain stage is marked by special symbols. The hollow circles indicate technologies that will reach mainstream adoption in two years.
- The gray circles represent technologies that will reach mainstream adoption in two to five years.
- The solid circles represent those that require five to 10 years to reach mainstream adoption, and the triangles denote those that require more than 10 years.
- The crossed circles represent technologies that will become obsolete before they reach the plateau.

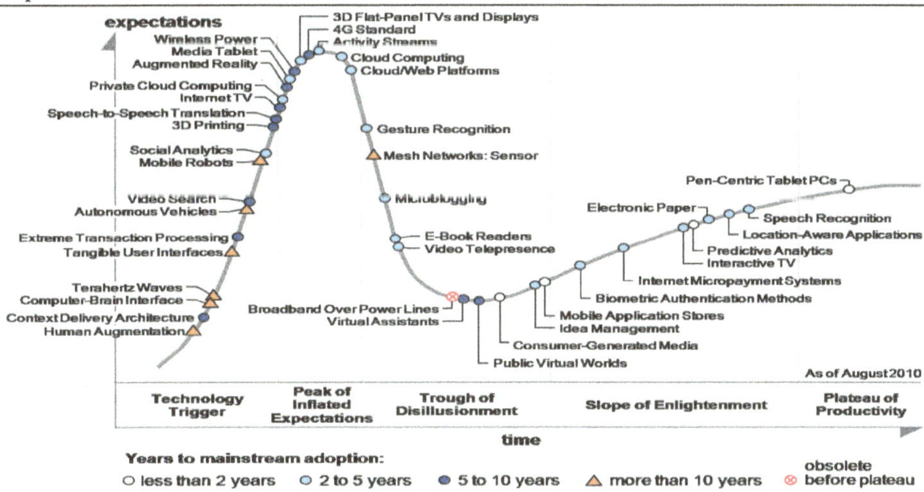

Figure 1.4 Hype cycle for Emerging Technologies, 2010

1.2.3 The Internet of Things and Cyber-Physical Systems

Internet of Things and the cyber physical systems are the evolutionary trends emphasize the extension of the Internet to everyday objects.

1.2.3.1 The Internet of Things

The concept of the IoT was introduced in 1999 at MIT.

- The IoT refers to the networked interconnection of everyday objects, tools, devices, or computers. The idea is to tag every object using RFID or a related sensor or electronic technology such as GPS.
- One can view the IoT as a wireless network of sensors that interconnect all things in our daily life. These things can be large or small and they vary with respect to time and place.
- The IoT needs to be designed to track 100 trillion static or moving objects simultaneously. The IoT demands universal addressability of all of the objects or things. To reduce the complexity of identification, search, and storage, one can set the threshold to filter out fine-grain objects.

Communication Pattern in IoT

Communication can be made between people and things or among the things themselves. Three communication patterns co-exist: namely

- H2H (human-to-human)
- H2T (human-tothing)
- and T2T (thing-to-thing)

Here things include machines such as PCs and mobile phones. The idea here is to connect things (including human and machine objects) at any time and any place intelligently with low cost. Any place connections include at the PC, indoor (away from PC), outdoors, and on the move. Any time connections include daytime, night, outdoors and indoors, and on the move as well.

- The dynamic connections will grow exponentially into a new dynamic network of networks, called the Internet of Things (IoT).
- Cloud computing researchers expect to use the cloud and future Internet technologies to support fast, efficient, and intelligent interactions among humans, machines, and any objects on Earth.
- A smart Earth should have intelligent cities, clean water, efficient power, convenient transportation, good food supplies, responsible banks, fast telecommunications, green IT, better schools, good health care, abundant resources, and so on.

1.2.3.2 Cyber-Physical Systems

- A cyber-physical system (CPS) is the result of interaction between computational processes and the physical world.
- A CPS integrates "cyber" (heterogeneous, asynchronous) with "physical" (concurrent and information-dense) objects.

- A CPS merges the "3C" technologies of computation, communication, and control into an intelligent closed feedback system between the physical world and the information world.
- CPS emphasizes exploration of virtual reality (VR) applications in the physical world.

1.3. Technologies for network based systems

In this we will focus on viable approaches to building distributed operating systems for handling massive parallelism in a distributed environment.

1.3.1 Multicore CPUs and Multithreading Technologies

The growth of component and network technologies over the past 30 years crucial to the development of HPC and HTC systems. In Figure 1.5, processor speed is measured in millions of instructions per second (MIPS) and network bandwidth is measured in megabits per second (Mbps) or gigabits per second (Gbps). The unit GE refers to 1 Gbps Ethernet bandwidth.

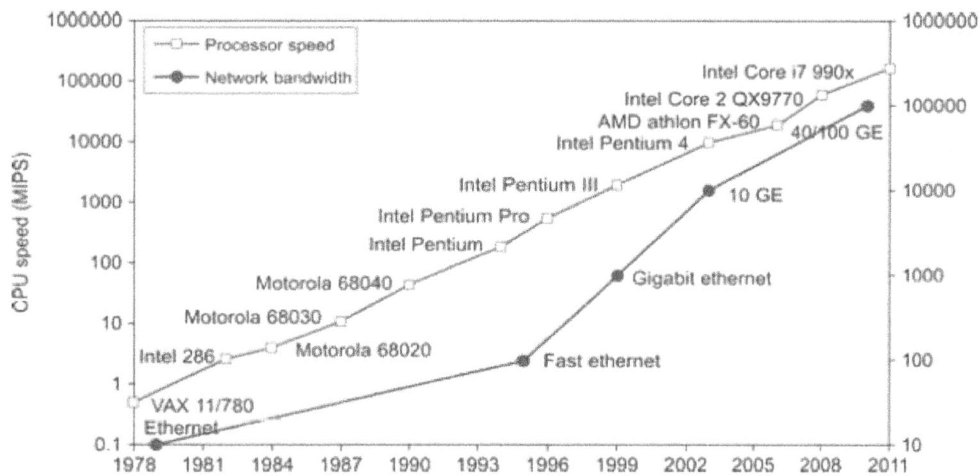

Figure 1.5 Improvement in processor and network technologies over 33 years

1.3.1.1 Advances in CPU Processors

- Today, advanced CPUs or microprocessor chips assume a multicore architecture with dual, quad, six, or more processing cores. These processors exploit parallelism at ILP and TLP levels.
- Processor speed growth is plotted in the upper curve in Figure 1.5 across generations of microprocessors or CMPs.
- The clock rate for these processors increased from 10 MHz for the Intel 286 to 4 GHz for the Pentium 4 in 30 years.
- The clock rate reached its limit on CMOS-based chips due to power limitations. Clock rate will not continue to improve unless chip technology matures.

- This limitation is attributed primarily to excessive heat generation with high frequency or high voltages.
- The ILP is highly exploited in modern CPU processors. ILP mechanisms include multiple-issue superscalar architecture, dynamic branch prediction, and speculative execution, among others. These ILP techniques demand hardware and compiler support.
- DLP and TLP are highly explored in graphics processing units (GPUs) that adopt many-core architecture with hundreds to thousands of simple cores.
- Both multi-core CPU and many-core GPU processors can handle multiple instruction threads at different magnitudes.

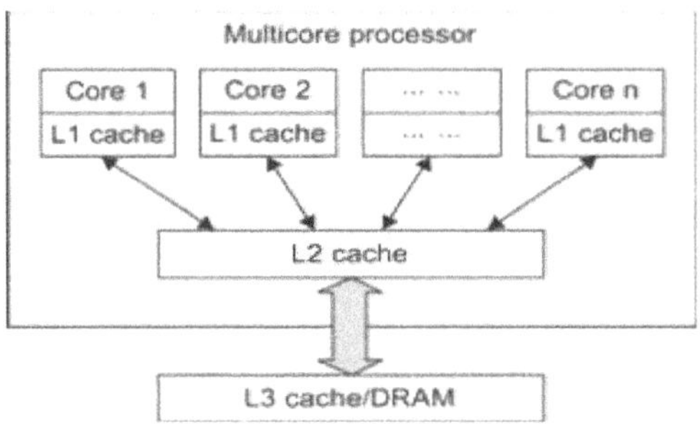

Figure 1.6 Schematic of a modern multicore CPU chip using a hierarchy of caches, where L1 cache is private to each core, on-chip L2 cache is shared and L3 cache or DRAM is off the chip.

Figure 1.6 shows the architecture of a typical multicore processor. Each core is essentially a processor with its own private cache (L1 cache). Multiple cores are housed in the same chip with an L2 cache that is shared by all cores. In the future, multiple CMPs could be built on the same CPU chip with even the L3 cache on the chip. Multicore and multithreaded CPUs are equipped with many high-end processors, including the Intel i7, Xeon, AMD Opteron, Sun Niagara, IBM Power 6, and X cell processors. Each core could be also multithreaded. For example, the Niagara II is built with eight cores with eight threads handled by each core. This implies that the maximum ILP and TLP that can be exploited in Niagara is 64 ($8 \times 8 = 64$).

1.3.1.2 Multicore CPU and Many-Core GPU Architectures

- Multicore CPUs may increase from the tens of cores to hundreds or more in the future. And the CPU has reached its limit in terms of exploiting massive DLP due to the abovementioned memory wall problem.
- This has triggered the development of many-core GPUs with hundreds or more thin cores. Both IA-32 and IA-64 instruction set architectures are built into commercial CPUs. Now, x-86 processors have been extended to serve HPC and HTC systems in some high-end server processors.
- Many RISC processors have been replaced with multicore x-86 processors and many-core GPUs in the Top 500 systems. This trend indicates that x-86 upgrades will dominate in data centers and supercomputers.

- The GPU also has been applied in large clusters to build supercomputers in MPPs.
- In the future, the processor industry is keen to develop asymmetric or heterogeneous chip multiprocessors that can house both fat CPU cores and thin GPU cores on the same chip.

1.3.1.3 Multithreading Technology

Consider in Figure 1.7 the dispatch of five independent threads of instructions to four pipelined data paths (functional units) in each of the following five processor categories, from left to right: a four issue superscalar processor, a fine-grain multithreaded processor, a coarse-grain multithreaded processor, a two-core CMP, and a simultaneous multithreaded (SMT) processor. The superscalar processor is single-threaded with four functional units. Each of the three multithreaded processors is four-way multithreaded over four functional data paths. In the dual-core processor, assume two processing cores, each a single-threaded two-way superscalar processor.

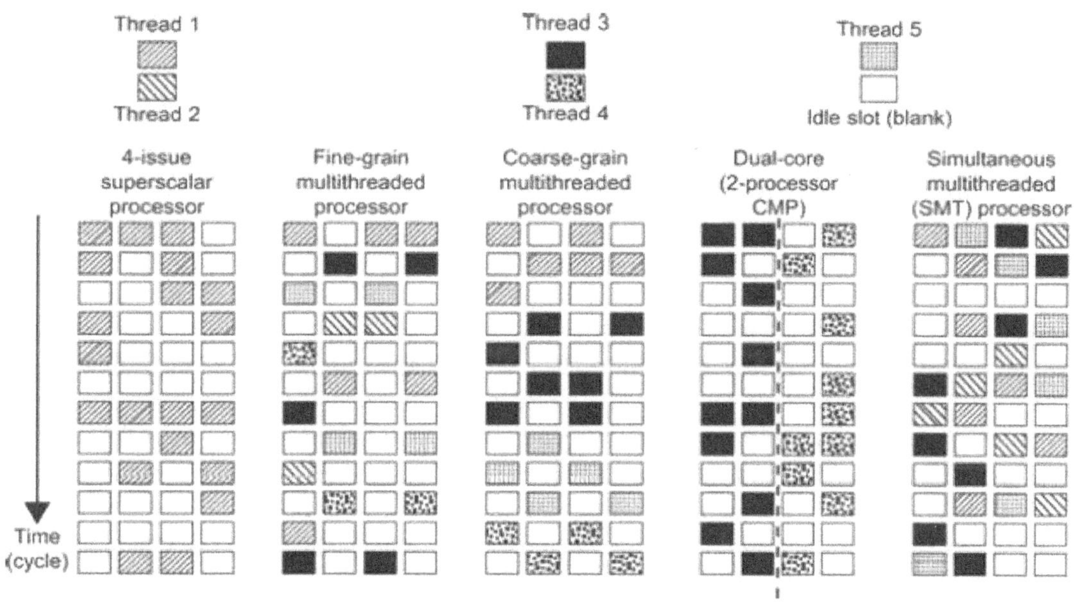

Figure 1.7 Five micro architectures in modern CPU processors that exploit ILP and TLP supported by multicore and multithreading technologies.

- Instructions from different threads are distinguished by specific shading patterns for instructions from five independent threads.
- Only instructions from the same thread are executed in a superscalar processor.
- Fine-grain multithreading switches the execution of instructions from different threads per cycle.
- Course-grain multithreading executes many instructions from the same thread for quite a few cycles before switching to another thread.
- The multicore CMP executes instructions from different threads completely.

- The SMT allows simultaneous scheduling of instructions from different threads in the same cycle.

These execution patterns closely mimic an ordinary program. The blank squares correspond to no available instructions for an instruction data path at a particular processor cycle. More blank cells imply lower scheduling efficiency. The maximum ILP or maximum TLP is difficult to achieve at each processor cycle.

1.3.2 GPU Computing to Exascale and Beyond

- A GPU is a graphics coprocessor or accelerator mounted on a computer's graphics card or video card. A GPU offloads the CPU from tedious graphics tasks in video editing applications.
- GPU chips can process a minimum of 10 million polygons per second, and are used in nearly every computer on the market today.
- Some GPU features were also integrated into certain CPUs. Traditional CPUs are structured with only a few cores. For example, the Xeon X5670 CPU has six cores. However, a modern GPU chip can be built with hundreds of processing cores.
- Unlike CPUs, GPUs have a throughput architecture that exploits massive parallelism by executing many concurrent threads slowly, instead of executing a single long thread in a conventional microprocessor very quickly.
- Lately, parallel GPUs or GPU clusters got lot of attention against the use of CPUs with limited parallelism. General-purpose computing on GPUs, known as GPGPUs, has appeared in the HPC field.

1.3.2.1 How GPUs Work

- Early GPUs functioned as coprocessors attached to the CPU.
- Each core on a GPU can handle eight threads of instructions. This translates to having up to 1,024 threads executed concurrently on a single GPU. This is true *massive parallelism*, compared to only a few threads that can be handled by a conventional CPU.
- The CPU is optimized for latency caches, while the GPU is optimized to deliver much *higher throughput* with explicit management of on-chip memory.
- Modern GPUs are *not restricted to accelerated graphics or video coding*. They are used in HPC systems to power supercomputers with massive parallelism at multicore and multithreading levels.
- GPUs are designed to handle large numbers of *floating-point operations in parallel.*
- The GPU *offloads the CPU* from all data-intensive calculations, not just those that are related to video processing.

Conventional GPUs are widely used in mobile phones, game consoles, embedded systems, PCs, and servers. The NVIDIA CUDA Tesla or Fermi is used in GPU clusters or in HPC systems for parallel processing of massive floating-pointing data.

1.3.2.2 GPU Programming Model

Figure 1.8 shows the interaction between a CPU and GPU in performing parallel execution of floating-point operations concurrently. The CPU is the conventional multicore processor with limited parallelism to exploit. The GPU has a many-core architecture that has hundreds of simple processing cores organized as multiprocessors. Each core can have one or more threads. Essentially, the CPU's floating-point kernel computation role is largely offloaded to the many-core GPU. The CPU instructs the GPU to perform massive data processing. The bandwidth must be matched between the on-board main memory and the on-chip GPU memory. This process is carried out in NVIDIA's CUDA programming using the GeForce 8800 or Tesla and Fermi GPUs.

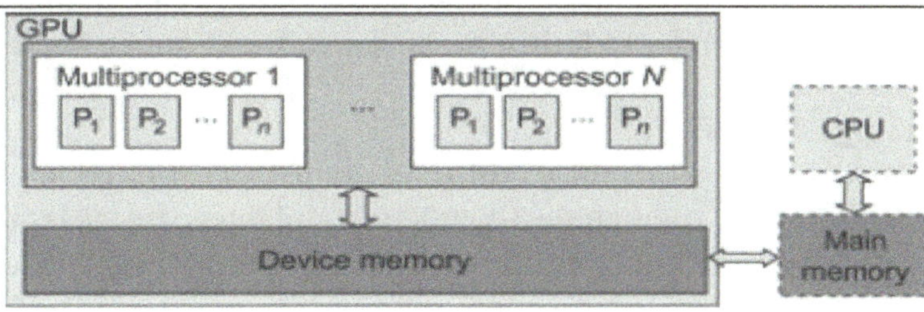

Figure 1.8 The use of a GPU along with a CPU for massively parallel execution in hundreds or thousands of processing cores

In the future, thousand-core GPUs may appear in Exascale (Eflops or 1018 flops) systems. This reflects a trend toward building future MPPs with hybrid architectures of both types of processing chips. In a DARPA report published in September 2008, four challenges are identified for Exascale computing: (1) energy and power, (2) memory and storage, (3) concurrency and locality, and (4) system resiliency.

1.3.2.3 Power Efficiency of the GPU

In the future Power and massive parallelism are the major benefits of GPUs over CPUs. By extrapolating current technology and computer architecture, it was estimated that 60 Gflops/watt per core is needed to run an exaflops system. Power constrains what we can put in a CPU or GPU chip. Dally has estimated that the CPU chip consumes about 2 nJ/instruction, while the GPU chip requires 200 pJ/instruction, which is 1/10 less than that of the CPU. The CPU is optimized for latency in caches and memory, while the GPU is optimized for throughput with explicit management of on-chip memory.

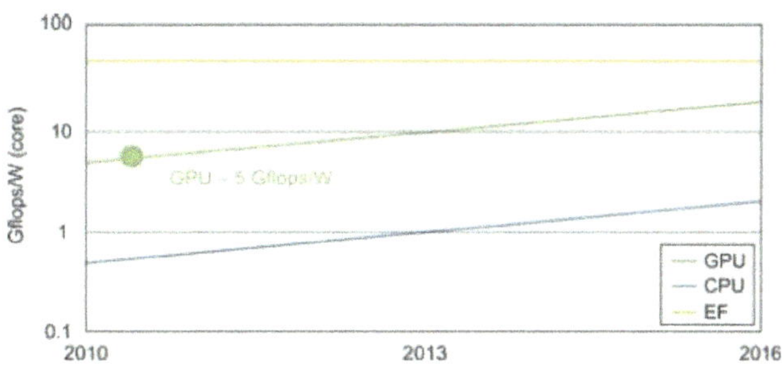

Figure 1.9 The GPU performance (middle line, measured 5 Gflops/W/core in 2011), compared with the lower CPU performance (lower line measured 0.8 Gflops/W/core in 2011) and the estimated 60 Gflops/W/core performance in 2011 for the Exascale (EF in upper curve) in the future

Figure 1.9 compares the CPU and GPU in their performance/power ratio measured in Gflops/watt per core. In 2010, the GPU had a value of 5 Gflops/watt at the core level, compared with less than 1 Gflop/watt per CPU core. This may limit the scaling of future supercomputers. However, the GPUs may close the gap with the CPUs. Data movement dominates power consumption. One needs to optimize the storage hierarchy and tailor the memory to the applications. We need to promote selfaware OS and runtime support and build locality-aware compilers and auto-tuners for GPU-based MPPs. This implies that both power and software are the real challenges in future parallel and distributed computing systems.

1.3.3.2 Disks and Storage Technology

The rapid growth of flash memory and solid-state drives (SSDs) are the future of HPC and HTC systems. A typical SSD can handle 300,000 to 1 million write cycles per block. SSD can last for several years, even under conditions of heavy write usage. Flash and SSD will demonstrate impressive speedups in many applications.

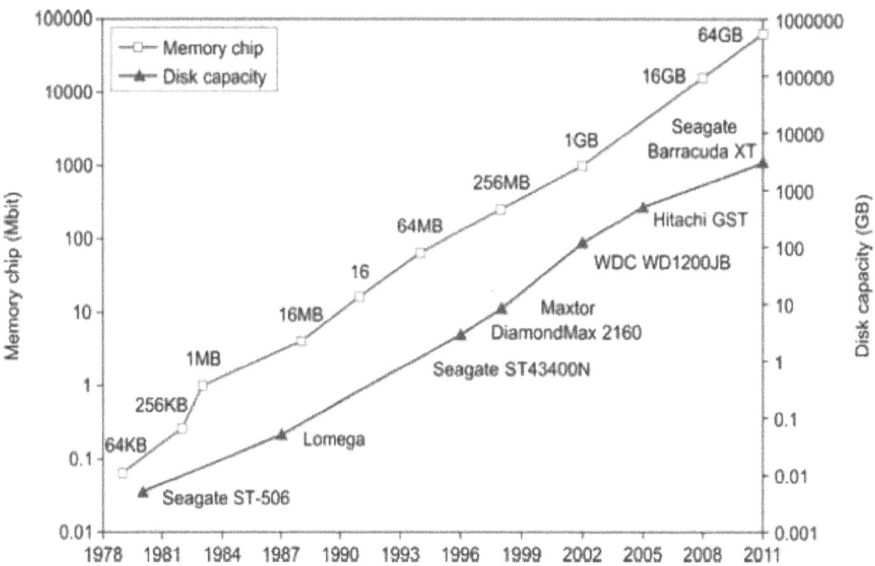

Figure 1.10 Improvement in memory and disk technologies over 33 years. The Seagate Barracuda XT disk has a capacity of 3 TB in 2011.

Eventually, power consumption, cooling, and packaging will limit large system development. Power increases linearly with respect to clock frequency and quadratic ally with respect to voltage applied on chips. Clock rate cannot be increased indefinitely. Lowered voltage supplies are very much in demand. In 2011, the SSDs are still too expensive to replace stable disk arrays in the storage market.

1.3.3.3 System-Area Interconnects

The nodes in small clusters are mostly interconnected by an Ethernet switch or a local area network (LAN). As Figure 1.11 shows, a LAN typically is used to connect client hosts to big servers. A storage area network (SAN) connects servers to network storage such as disk arrays. Network attached storage (NAS) connects client hosts directly to the disk arrays. All three types of networks often appear in a large cluster built with commercial network components. If no large distributed storage is shared, a small cluster could be built with a multiport Gigabit Ethernet switch plus copper cables to link the end machines. All three types of networks are commercially available.

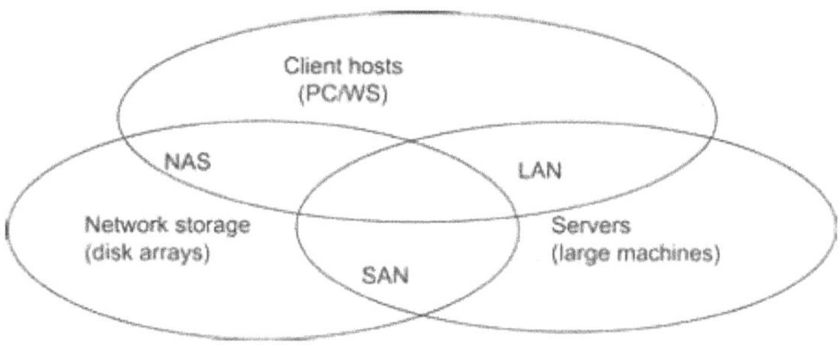

Figure 1.11 Three interconnection networks for connecting servers, client hosts, and storage devices; the LAN connects client hosts and servers, the SAN connects servers with disk arrays, and the NAS connects clients with large storage systems in the network environment.

1.3.3.4 Wide-Area Networking

The lower curve in Figure 1.10 plots the rapid growth of Ethernet bandwidth from 10 Mbps in 1979 to 1 Gbps in 1999 and 40 ~ 100 GE in 2011. It has been speculated that 1 Tbps network links will become available by 2013. An increase factor of two per year on network performance was reported, which is faster than Moore's law on CPU speed doubling every 18 months. The implication is that more computers will be used concurrently in the future. High-bandwidth networking increases the capability of building massively distributed systems. The IDC 2010 report predicted that both InfiniBand and Ethernet will be the two major interconnect choices in the HPC arena. Most data centers are using Gigabit Ethernet as the interconnect in their server clusters.

1.3.4 Virtual Machines and Virtualization Middleware

- A conventional computer has a single OS image. This offers a rigid architecture that tightly couples application software to a specific hardware platform. Some software running well on one machine may not be executable on another platform with a different instruction set under a fixed OS.
- Virtual machines (VMs) offer novel solutions to underutilized resources, application inflexibility, software manageability, and security concerns in existing physical machines.
- To build large clusters, grids, and clouds, need to access large amounts of computing, storage, and networking resources in a virtualized manner.
- Virtualization aggregates those resources, and hopefully, offers a single system image.

Figure 1.12 illustrates the architectures of three VM configurations.

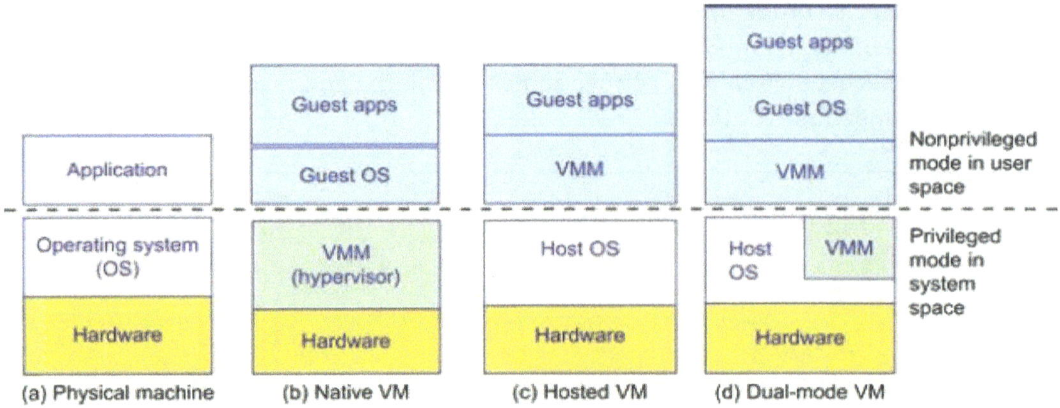

Figure 1.12 Three VM architectures in (b), (c), and (d), compared with the traditional physical machine shown in (a).

1.4.4.1 Virtual Machines

- The VM can be provisioned for any hardware system.
- The VM is built with virtual resources managed by a guest OS to run a specific application.
- Between the VMs and the host platform, one needs to deploy a middleware layer called a virtual machine monitor (VMM).
- The VM approach offers hardware independence of the OS and applications. The user application running on its dedicated OS could be bundled together as a virtual appliance that can be ported to any hardware platform.
- The VM could run on an OS different from that of the host computer.

Hypervisor approach- different architectures:

- **Bare-metal VM**- the hypervisor handles the bare hardware (CPU, memory, and I/O) directly.
  For example the hardware has x-86 architecture running the Windows system. The guest OS could be a Linux system and the hypervisor is the XEN system.
- **Host VM** -VMM runs in non privileged mode. The host OS need not be modified.
- **Dual mode VM** - part of the VMM runs at the user level and another part runs at the supervisor level. In this case, the host OS may have to be modified to some extent.
- **Multiple VMs** can be ported to a given hardware system to support the virtualization process.

1.4.4.2 VM Primitive Operations

The VMM provides the VM abstraction to the guest OS. With full virtualization, the VMM exports a VM abstraction identical to the physical machine.

- First, the VMs can be *multiplexed* between hardware machines.
- Second, a VM can be *suspended* and stored in stable storage.

- Third, a suspended VM can be *resumed or provisioned* to a new hardware platform.
- Finally, a VM can be *migrated* from one hardware platform to another.

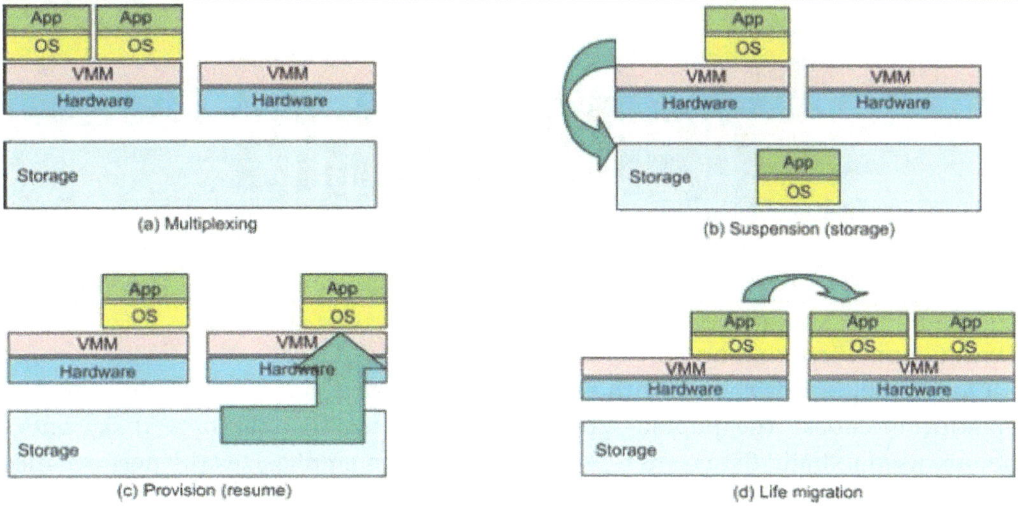

Figure 1.13 VM multiplexing, suspension, provision, and migration in a distributed computing environment

Advantages of VM approach:

- These VM operations enable a VM to be provisioned to any available hardware platform.
- They enable flexibility in porting distributed application executions.
- The VM approach will significantly enhance the utilization of server resources.
- Multiple server functions can be consolidated on the same hardware platform to achieve higher system efficiency.

1.4.4.3 Virtual Infrastructures

- Physical resources to compute, storage, and networking (Figure 1.14) are mapped to the needy applications embedded in various VMs at the top.
- Hardware and software are then separated.
- Virtual infrastructure connects resources to distributed applications.
- Dynamic mapping of system resources to specific applications results in decreased costs and increased efficiency and responsiveness.
- Virtualization for server consolidation and containment is a good example of this.

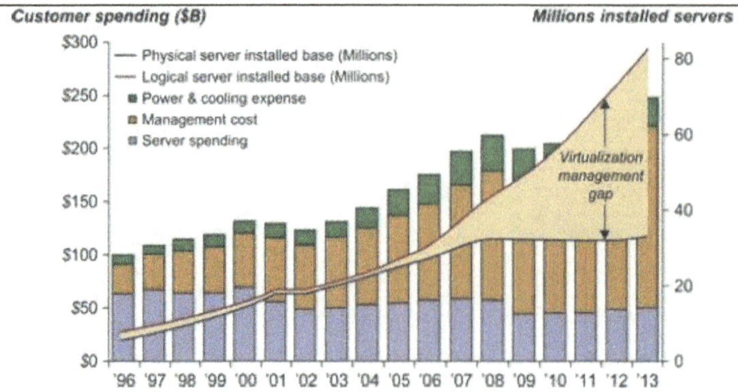

Figure 1.14 Growth and cost breakdown of data centers over the years.

1.4.5 Data Center Virtualization for Cloud Computing

Cloud architecture is built with commodity hardware and network devices. Almost all cloud platforms choose the popular x86 processors. Low-cost terabyte disks and Gigabit Ethernet are used to build data centers. Data center design emphasizes the performance/price ratio over speed performance alone. In other words, storage and energy efficiency are more important than shear speed performance.

1.4.5.1 Data Center Growth and Cost Breakdown

- **Data Center Growth**
  - ✓ A *large data center* may be built with thousands of servers and *smaller data centers* are typically built with hundreds of servers.
  - ✓ The cost to build and maintain data center servers has increased over the years.
- **Cost Break down**
  - ✓ 30 percent of data center costs goes toward purchasing servers and disks
  - ✓ 33 percent is attributed to the chiller
  - ✓ 18 percent to the uninterruptible power supply (UPS)
  - ✓ 9 percent to computer room air conditioning (CRAC)
  - ✓ 7 percent to power distribution, lighting, and transformer costs.

Thus, about 60 percent of the cost to run a data center is allocated to management and maintenance. The server purchase cost did not increase much with time. The cost of electricity and cooling did increase from 5 percent to 14 percent in 15 years.

1.4.5.2 Low-Cost Design Philosophy

- *High-end switches or routers* may be too cost-prohibitive for building data centers. Thus, using high bandwidth networks may not fit the economics of cloud computing.
- *Given a fixed budget*, commodity switches and networks are more desirable in data centers.
- The *software layer* handles network traffic balancing, fault tolerance, and expandability.

1.4.5.3 Convergence of Technologies

Essentially, cloud computing is enabled by the convergence of technologies in four areas:

(1) **Hardware virtualization and multi-core chips**- enable the existence of dynamic configurations in the cloud.

(2) **Utility and grid computing** - technologies lay the necessary foundation for computing clouds.

(3) **SOA, Web 2.0, and WS mashups**-of platforms are pushing the cloud another step forward

(4) **Atonomic computing and data center automation**-operations contribute to the rise of cloud computing.

How to manage and analyze information?

- Data comes from sensors, lab experiments, simulations, individual archives, and the web in all scales and formats. Preservation, movement, and access of massive data sets require generic tools supporting high-performance, scalable file systems, databases, algorithms, workflows, and visualization.
- With science becoming data-centric, a new paradigm of scientific discovery is becoming based on data-intensive technologies.

The Computer Science and Telecommunication Board (CSTB) recommended fostering tools for data capture, data creation, and data analysis. A cycle of interaction exists among four technical areas.

- First, cloud technology is driven by a surge of interest in data deluge.
  - ✓ Cloud computing impacts e-science greatly, which explores multicore and parallel computing technologies. These two hot areas enable the buildup of data deluge.
  - ✓ To support data-intensive computing, one needs to address workflows, databases, algorithms, and virtualization issues.
- Second, by linking computer science and technologies with scientists, a spectrum of e-science or e-research applications in biology, chemistry, physics, the social sciences, and the humanities has generated new insights from interdisciplinary activities.
  - ✓ Cloud computing is a transformative approach as it promises much more than a data center model. It fundamentally changes how we interact with information. The cloud provides services on demand at the infrastructure, platform, or software level.
- Third, at the platform level, MapReduce offers a new programming model that transparently handles data parallelism with natural fault tolerance capability.
  - ✓ Iterative MapReduce extends MapReduce to support a broader range of data mining algorithms commonly used in scientific applications.
  - ✓ The cloud runs on an extremely large cluster of commodity computers. Internal to each cluster node, multithreading is practiced with a large number of cores in many-core GPU clusters.
- Finally, Data-intensive science, cloud computing, and multicore computing are converging and revolutionizing the next generation of computing in architectural design and programming challenges. They enable the pipeline: Data becomes information and knowledge, and in turn becomes machine wisdom as desired in SOA.

1.5 Clusters of Cooperative Computers

A computing cluster consists of interconnected stand-alone computers which work cooperatively as a single integrated computing resource.

1.5.1 Cluster Architecture

Figure 1.15 shows the architecture of a typical server cluster built around a low-latency, high bandwidth interconnection network.

- This network can be as simple as a SAN (e.g., Myrinet) or a LAN (e.g., Ethernet).
- To build a larger cluster with more nodes, the interconnection network can be built with multiple levels of Gigabit Ethernet, Myrinet, or InfiniBand switches.
- Through hierarchical construction using a SAN, LAN, or WAN, one can build scalable clusters with an increasing number of nodes. The cluster is connected to the Internet via a virtual private network (VPN) gateway.
- The gateway IP address locates the cluster. The system image of a computer is decided by the way the OS manages the shared cluster resources.
- Most clusters have loosely coupled node computers.
- All resources of a server node are managed by their own OS. Thus, most clusters have multiple system images as a result of having many autonomous nodes under different OS control.

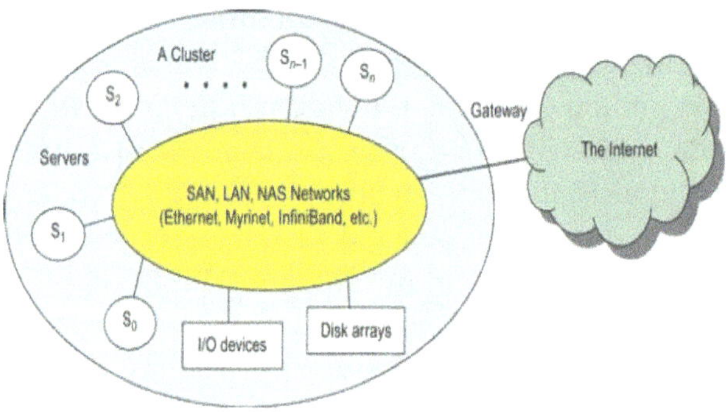

Figure 1.15 A cluster of servers interconnected by a high-bandwidth SAN or LAN with shared I/O devices and disk arrays; the cluster acts as a single computer attached to the Internet.

1.5.2 Single-System Image

- An ideal cluster should merge multiple system images into a single-system image (SSI).
- Cluster designers desire a cluster operating system or some middleware to support SSI at various levels, including the sharing of CPUs, memory, and I/O across all cluster nodes.
- An SSI is an illusion created by software or hardware that presents a collection of resources as one integrated, powerful resource.
- SSI makes the cluster appear like a single machine to the user.

- A cluster with multiple system images is nothing but a collection of independent computers.

1.5.3 Hardware, Software, and Middleware Support

Hardware

- Clusters exploring massive parallelism are commonly known as MPPs.
- The building blocks are computer nodes (PCs, workstations, servers, or SMP), special communication software such as PVM or MPI, and a network interface card in each computer node.

Software

- Most clusters run under the Linux OS.
- The computer nodes are interconnected by a high-bandwidth network (such as Gigabit Ethernet, Myrinet, InfiniBand, etc.).

Middleware

- Special cluster middleware supports are needed to create SSI or high availability (HA).
- Both sequential and parallel applications can run on the cluster, and special parallel environments are needed to facilitate use of the cluster resources.

For example, distributed memory has multiple images. Users may want all distributed memory to be shared by all servers by forming distributed shared memory (DSM). Many SSI features are expensive or difficult to achieve at various cluster operational levels. Instead of achieving SSI, many clusters are loosely coupled machines. Using virtualization, one can build many virtual clusters dynamically, upon user demand.

1.5.4 Major Cluster Design Issues

- A cluster-wide OS for complete *resource sharing* is not available yet.
- *Middleware or OS extensions* were developed at the user space to achieve SSI at selected functional levels. Without this middleware, cluster nodes cannot work together effectively to *achieve cooperative computing*.
- The *software environments and applications* must rely on the middleware to achieve high performance.
- The cluster benefits come from scalable *performance, efficient message passing, high system availability, seamless fault tolerance, and cluster-wide job management*, as summarized in Table 1.2.

Table 1.2 Critical Cluster Design Issues and Feasible Implementations

Features	Functional Characterization	Feasible Implementations
Availability and Support	Hardware and software support for sustained HA in cluster	Failover, failback, check pointing, rollback recovery, nonstop OS, etc.
Hardware Fault Tolerance	Automated failure management to eliminate all single points of failure	Component redundancy, hot swapping, RAID, multiple power supplies, etc.
Single System Image (SSI)	Achieving SSI at functional level with hardware and software support, middleware, or OS extensions	Hardware mechanisms or middleware support to achieve DSM at coherent cache level
Efficient Communications	To reduce message-passing system overhead and hide latencies	Fast message passing, active messages, enhanced MPI library, etc.

Cluster-wide Job Management	Using a global job management system with better scheduling and monitoring	Application of single-job management systems such as LSF, Codine, etc.
Dynamic Load Balancing	Balancing the workload of all processing nodes along with failure recovery	Workload monitoring, process migration, job replication and gang scheduling, etc.
Scalability and Programmability	Adding more servers to a cluster or adding more clusters to a grid as the workload or data set increases	Use of scalable interconnect, performance monitoring, distributed execution environment, and better software tools

1.6. Grid Computing Infrastructures

In the past 30 years, users have experienced a natural growth path from Internet to web and grid computing services. Internet services such as the Telnet command enables a local computer to connect to a remote computer. A web service such as HTTP enables remote access of remote web pages. Grid computing is envisioned to allow close interaction among applications running on distant computers simultaneously.

1.6.1 Computational Grids

- A computing grid offers an infrastructure that couples computers, software/middleware, special instruments, and people and sensors together.
- The grid is often constructed across LAN, WAN, or Internet backbone networks at a regional, national, or global scale.
- Enterprises or organizations present grids as integrated computing resources. They can also be viewed as virtual platforms to support virtual organizations.
- The computers used in a grid are primarily workstations, servers, clusters, and supercomputers. Personal computers, laptops, and PDAs can be used as access devices to a grid system.

Figure 1.16 shows an example computational grid built over multiple resource sites owned by different organizations. The resource sites offer complementary computing resources, including workstations, large servers, a mesh of processors, and Linux clusters to satisfy a chain of computational needs. The grid is built across various IP broadband networks including LANs and WANs already used by enterprises or organizations over the Internet. The grid is presented to users as an integrated resource pools as shown in the upper half of the figure.

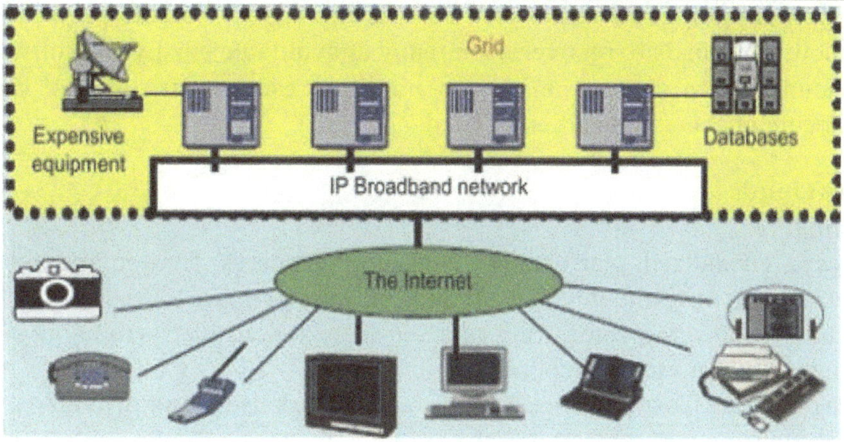

Figure 1.16 Computational grid or data grid providing computing utility, data, and information services through resource sharing and cooperation among participating organizations.

At the server end, the grid is a network. At the client end, we see wired or wireless terminal devices. The grid integrates the computing, communication, contents, and transactions as rented services. Enterprises and consumers form the user base, which then defines the usage trends and service characteristics.

1.6.2 Grid Families

Grid technology demands new distributed computing models, software/middleware support, network protocols, and hardware infrastructures. In Table 1.3, grid systems are classified in essentially two categories: *computational or data grids and P2P grids.*

Design Issues	Computational and Data Grids	P2P Grids
Grid Applications Reported	Distributed supercomputing, National Grid initiatives, etc.	Open grid with P2P flexibility, all resources from client machines
Representative Systems	TeraGrid built in US, ChinaGrid in China, and the e-Science grid built in UK	JXTA,FightAid@home, SETI@home
Development Lessons Learned	Restricted user groups, middleware bugs, protocols to acquire resources	Unreliable user-contributed resources, limited to a few apps

Table 1.4 Two Grid Computing Infrastructures and Representative Systems

1.7 Cloud Computing Over the Internet

Cloud computing has been defined as "A cloud is a pool of virtualized computer resources. A cloud can host a variety of different workloads, including batch-style backend jobs and interactive and user-facing applications."

- A cloud allows workloads to be deployed and scaled out quickly through rapid provisioning of virtual or physical machines.

- The cloud supports redundant, self-recovering, highly scalable programming models that allow workloads to recover from many unavoidable hardware/software failures.
- The cloud system should be able to monitor resource use in real time to enable rebalancing of allocations when needed.

1.7.1 Internet Clouds

Cloud computing

➔ Applies a virtualized platform with elastic resources on demand by provisioning hardware, software, and data sets dynamically (see Figure 1.18)

➔ Idea to move desktop computing to a service-oriented platform using server clusters and huge databases at data centers.

➔ Leverages its low cost and simplicity to benefit both users and providers.

➔ Enables Machine virtualization with cost-effectiveness.

➔ Intends to satisfy many user applications simultaneously.

➔ Ecosystem must be designed to be secure, trustworthy, and dependable.

Some computer users think of the cloud as a centralized resource pool. Others consider the cloud to be a server cluster which practices distributed computing over all the servers used.

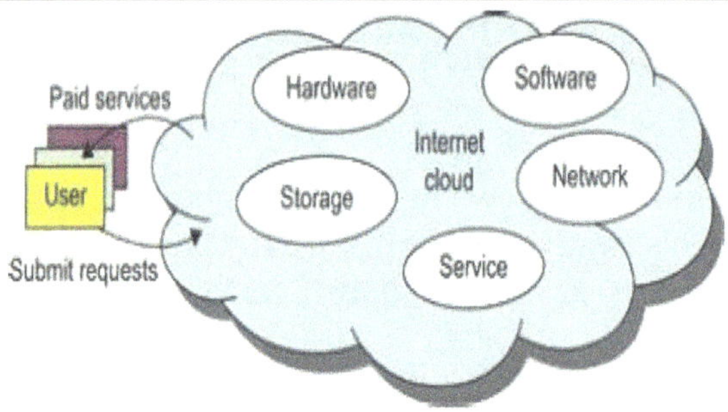

Figure 1.18 Virtualized resources from data centers to form an Internet cloud, provisioned with hardware, software, storage, network, and services for paid users to run their applications.

1.7.2 The Cloud Landscape

A distributed computing system tends to be owned and operated by an autonomous administrative domain for on-premises computing needs. These traditional systems have encountered several performance bottlenecks:

➔ constant system maintenance

➔ poor utilization

➔ and increasing costs associated with hardware/software upgrades

Cloud computing as an on-demand computing paradigm resolves or relieves us from these problems.

Figure 1.19 depicts the cloud landscape and major cloud players, based on three cloud service models.

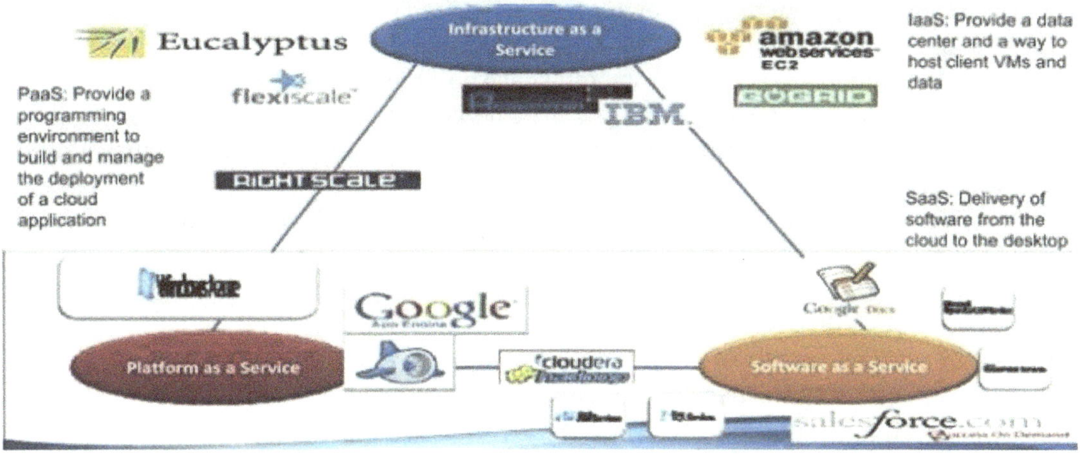

Figure 1.19 Three cloud service models in a cloud landscape of major providers.

Infrastructure as a Service (IaaS)

- This model puts together infrastructures demanded by users namely servers, storage, networks, and the data center fabric.
- The user can deploy and run on multiple VMs running guest OSes on specific applications.
- The user does not manage or control the underlying cloud infrastructure, but can specify when to request and release the needed resources.

Platform as a Service (PaaS)

- This model enables the user to deploy user-built applications onto a virtualized cloud platform.
- PaaS includes middleware, databases, development tools, and some runtime support such as Web 2.0 and Java.
- The platform includes both hardware and software integrated with specific programming interfaces.
- The provider supplies the API and software tools. The user is freed from managing the cloud infrastructure.

Software as a Service (SaaS)

- This refers to browser-initiated application software over thousands of paid cloud customers.
- The SaaS model applies to business processes, industry applications, consumer relationship management (CRM), enterprise resources planning (ERP), human resources (HR), and collaborative applications.
- On the customer side, there is no upfront investment in servers or software licensing.
- On the provider side, costs are rather low, compared with conventional hosting of user applications.

Internet clouds offer four deployment modes: *private, public, managed, and hybrid*. These modes demand different levels of security implications. The different SLAs imply that the security responsibility is shared among all the cloud providers, the cloud resource consumers, and the third party cloud-enabled software providers.

The following list highlights eight reasons to adapt the cloud for upgraded Internet applications and web services:

1. Desired location in areas with protected space and higher energy efficiency
2. Sharing of peak-load capacity among a large pool of users, improving overall utilization
3. Separation of infrastructure maintenance duties from domain-specific application development
4. Significant reduction in cloud computing cost, compared with traditional computing paradigms
5. Cloud computing programming and application development
6. Service and data discovery and content/service distribution
7. Privacy, security, copyright, and reliability issues
8. Service agreements, business models, and pricing policies

1.8 Software Environments for Distributed Systems and Clouds

This section introduces popular software environments for using distributed and cloud computing systems.

1.8.1 Service-Oriented Architecture (SOA)

- In grids/web services, Java, and CORBA, an entity is,
  - ➔ a service,
  - ➔ a Java object, and
  - ➔ a CORBA distributed object.
- The architectures build on the OSI layers that provide the base networking abstractions.
- On top of this a *base software environment* .NET or Apache Axis for web services, the Java Virtual Machine for Java, and a broker network for CORBA are present.
- On top of this base environment one would *build a higher level environment* reflecting the special features of the distributed computing environment.
- This starts with entity interfaces and inter-entity communication, which rebuild the top four OSI layers but at the entity and not the bit level.

Figure 1.20 shows the layered architecture for distributed entities used in web services and grid systems.

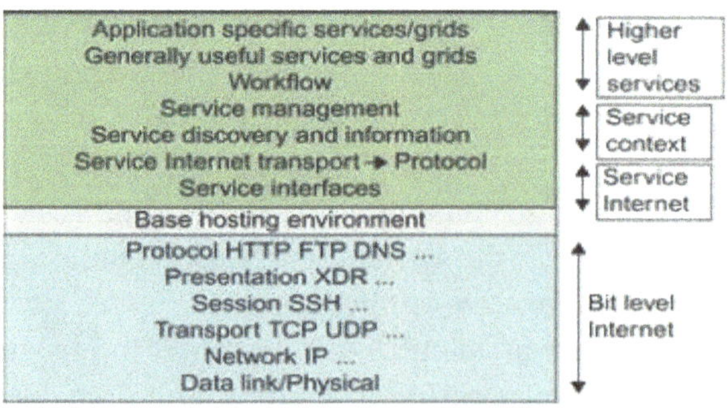

Figure 1.20 Layered architecture for web services and the grids

1.8.1.1 Layered Architecture for Web Services and Grids

Layered Architecture

- The **Entity interfaces** correspond to the Web Services Description Language (WSDL), Java method, and CORBA interface definition language (IDL) specifications in distributed systems.
  - These interfaces are linked with **customized, high-level communication** systems in SOAP, RMI, and IIOP
  - These communication systems **support features** including particular message patterns (such as Remote Procedure Call or RPC), fault recovery, and specialized routing.
  - These communication systems are **built on** message-oriented middleware (enterprise bus) infrastructure such as WebSphere MQ or Java Message Service (JMS) which provide rich functionality and support virtualization of routing, senders, and recipients.
- The **fault tolerance**, the features in the Web Services Reliable Messaging (WSRM) framework mimic the OSI layer capability (as in TCP fault tolerance) modified to match the different abstractions (such as messages versus packets, virtualized addressing) at the entity levels.
- **Security** is a critical capability that uses the concepts of Internet Protocol Security (IPsec) and secure sockets in the OSI layers. Entity communication is supported by higher level services for registries, metadata, and management of the entities.

Web Services and Grids

- JNDI (Jini and Java Naming and Directory Interface) illustrating different approaches within the Java distributed object model.
- The CORBA Trading Service, UDDI (Universal Description, Discovery, and Integration), LDAP (Lightweight Directory Access Protocol), and ebXML (Electronic Business using eXtensible Markup Language) are other examples of **discovery and information services.**
- **Management services** include service state and lifetime support; examples include the CORBA Life Cycle and Persistent states, the different Enterprise JavaBeans models, Jini's lifetime model, and a suite of web services specifications.

The distributed model has the following advantages:

- offers a "shared memory" model allowing more convenient exchange of information.
- higher performance (from multiple CPUs when communication is unimportant).
- Separation of software functions with software reuse and maintenance.

1.8.1.2 Web Services and Tools

Two choices of service architecture:

- Web services
- REST systems

Both web services and REST systems have very distinct approaches to building reliable interoperable systems.

Web Service

- Fully specify all aspects of the service and its environment.
- Specification is carried with communicated messages using Simple Object Access Protocol (SOAP).
- The hosting environment becomes a universal distributed operating system with fully distributed capability carried by SOAP messages.
- This approach has been hard to agree on key parts of the protocol and even harder to efficiently implement the protocol by software such as Apache Axis.

REST Systems

- Simplicity and delegates most of the difficult problems to application (implementation-specific) software.
- REST has minimal information in the header, and the message body carries all the needed information.
- REST architectures are clearly more appropriate for rapid technology environments.
- REST can use XML schemas but not those that are part of SOAP; "XML over HTTP"
- Above the communication and management layers, REST has the ability to compose new entities or distributed programs by integrating several entities together.

Tools

- ➢ In CORBA and Java, the distributed entities are linked with RPCs,
  - ✓ The simplest way to build composite applications is to view the entities as objects and use the traditional ways of linking them together.
- ➢ Java, this could be as simple as writing a Java program with method calls replaced by Remote Method Invocation (RMI),
- ➢ CORBA supports a similar model with a syntax reflecting the C++ style of its entity (object) interfaces.
- ➢ The term "grid" to refer to a single service or a collection of services, sensors represent entities that output data (as messages), and grids and clouds represent collections of services that have multiple message-based inputs and outputs.

1.8.1.3 The Evolution of SOA

Service-Oriented Architecture (SOA)

- As shown in Figure 1.21, service-oriented architecture (SOA) has evolved over the years.
- SOA applies to building grids, clouds, grids of clouds, clouds of grids, clouds of clouds (also known as inter clouds), and systems of systems in general.
- A large number of sensors provide data-collection services, denoted in the figure as SS (sensor service).
- A sensor can be a ZigBee device, a Bluetooth device, a WiFi access point, a personal computer, a GPA, or a wireless phone, among other things.
- Raw data is collected by sensor services.

- All the SS devices interact with large or small computers, many forms of grids, databases, the compute cloud, the storage cloud, the filter cloud, the discovery cloud, and so on.
- Filter services (fs in the figure) are used to eliminate unwanted raw data, in order to respond to specific requests from the web, the grid, or web services.
- A collection of filter services forms a filter cloud.
- SOA aims to search for, or sort out, the useful data from the massive amounts of raw data items. Processing this data will generate useful information, and subsequently, the knowledge for our daily use. In fact, wisdom or intelligence is sorted out of large knowledge bases.

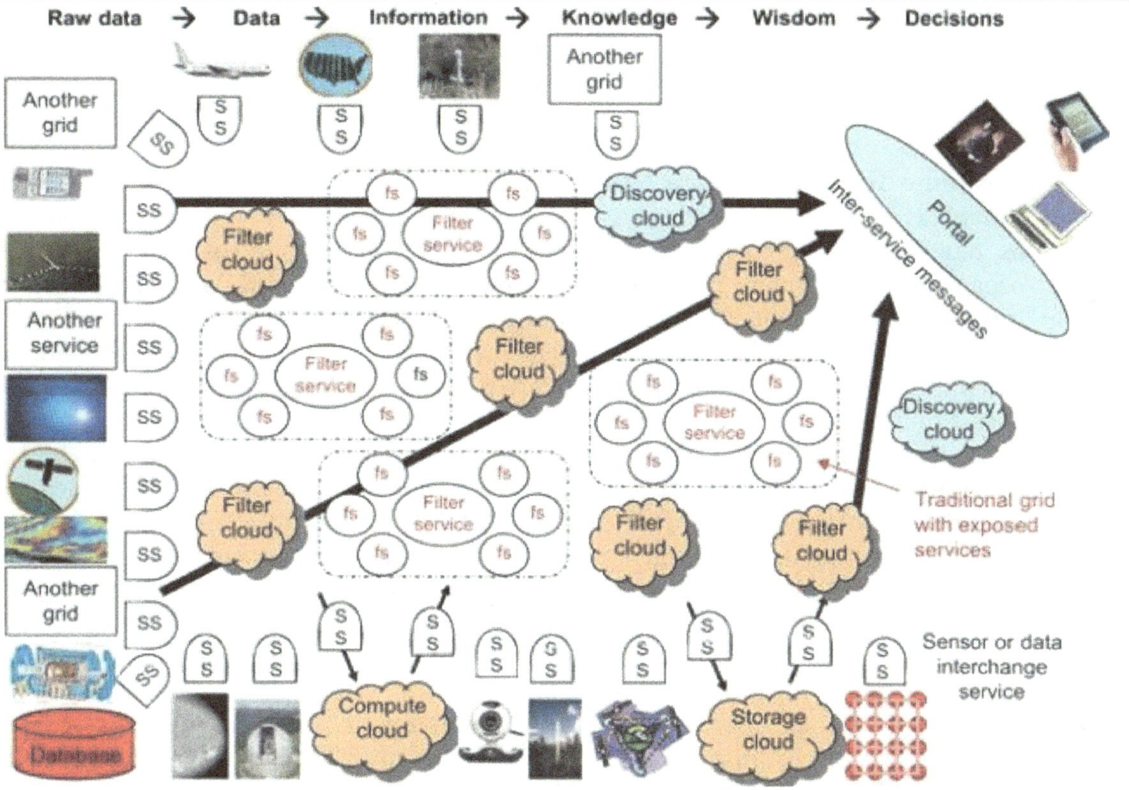

Figure 1.21 The evolution of SOA: grids of clouds and grids, where "SS" refers to a sensor service and "fs" to a filter or transforming service.

Most distributed systems require a web interface or portal. For raw data collected by a large number of sensors to be transformed into useful information or knowledge, the data stream may go through a sequence of compute, storage, filter, and discovery clouds. Finally, the inter-service messages converge at the portal, which is accessed by all users. Two example portals are OGFCE and HUBzero.

1.8.1.3 Grids versus Clouds

- ➔ In general, a grid system applies static resources, while a cloud emphasizes elastic resources.
- ➔ The differences between grids and clouds are limited only in dynamic resource allocation based on virtualization and autonomic computing.
- ➔ One can build a grid out of multiple clouds. This type of grid can do a better job than a pure cloud, because it can explicitly support negotiated resource allocation.
- ➔ Thus one may end up building with a system of systems: such as a cloud of clouds, a grid of clouds, or a cloud of grids, or inter-clouds as a basic SOA architecture.

Introduction to Grid Architecture and standards

1.9 Introduction to Grid

Grid computing has emerged as an important field synonymous to high throughput computing (HTC). The importance of grids is defined in terms of the amount of work they are able to deliver over a period of time.

Definition of Grid

[1] According to IBM's definition "grid is a collection of distributed computing resources available over a local or wide area network that appear to an end user or application as one large virtual computing system. The vision is to create virtual dynamic organizations through secure, coordinated resource-sharing among individuals, institutions, and resources. Grid computing is an approach to distributed computing that spans not only locations but also organizations, machine architectures, and software boundaries to provide unlimited power, collaboration, and information access to everyone connected to a grid."

[2] According to the Globus Alliance "The grid refers to an infrastructure that enables the integrated, collaborative use of high-end computers, networks, databases, and scientific instruments owned and managed by multiple organizations. Grid applications often involve large amounts of data and/or computing and often require secure resource sharing across organizational boundaries, and are thus not easily handled by today's Internet and Web infrastructures."

[3] Industry-formulated definition of grid computing is "A computational grid is a hardware and software infrastructure that provides dependable, consistent, pervasive, and inexpensive access to high-end computational capabilities. A grid is concerned with coordinated resource sharing and problem solving in dynamic, multi-institutional virtual organizations. The key concept is the ability to negotiate resource-sharing arrangements among a set of participating parties (providers and consumers) and then to use the resulting resource pool for some purpose. The sharing that we are concerned with is not primarily file exchange but rather direct access to computers, software, data, and other resources, as is required by a range of collaborative problem-solving and resource-brokering strategies emerging in industry, science, and engineering. This sharing is, necessarily, highly controlled, with resource providers and consumers defining clearly and carefully just what is shared, who is allowed to share, and the conditions under which sharing occurs. A set of individuals and/or institutions defined by such sharing rules form what we call a virtual organization (VO)."

Grids from Existing Technologies

- Grids have evolved from existing technologies such as distributed computing, web services, the Internet, various cryptography technologies providing security features and virtualization technology.
- The grid technology takes features from these technologies to develop a system that can provide computational resources for some specific tasks.
- These tasks can be the simulation of stock markets to predict future trends, scientific research such as prediction of earthquakes or serving business needs for an organization having a geographically distributed presence.
- Grid is an evolutionary technology, which leverages existing IT, infrastructure to provide high throughput computing.

Virtualization

- Virtualization in grids refers to seamless integration of geographically distributed and heterogeneous systems.
- This enables users to make use of the services provided by the grid in a transparent way the users need not be aware of the location of computing resources.
- So, from the users' perspective, there is just one point of entry to the grid system. They just have to submit their service request at this node.

Virtual Organization

- Virtual Organization in grid is defined as a "dynamic collection of multiple organizations providing coordinated resource sharing".
- The formation of VO is aimed at utilizing computing resources for specific problem solving .
- Based on the concept of VOs, we review three terms
  - ➔ The first terms are **virtualization**, which has already been explained and stems from virtual organizations.
  - ➔ The second term is **heterogeneity** is a multi-institutional entity. The organizations that form part of a VO may have different resources in terms of hardware, operating system and network bandwidth. So, we infer that a VO is a collection of heterogeneous resources.
  - ➔ The third term is **dynamic**. Organizations can join or leave a VO per their requirements and convenience. So a VO is a dynamic entity.
- According to it, a grid should provide resource coordination minus centralized control, it should be based on open standards, and it should provide a nontrivial quality of service. A grid can be used for computational purposes (computational grid), for storage of data on a large scale (data grid), or a combination of both.

1.9.1 Grid versus Other Distributed Systems

The major differences between grid and other distributed systems based on Remote Method Invocation (RMI) and Common Object Request Broker Architecture.

Distributed Systems	Grid
• Serve the purpose of a single organization and have a centralized control	• Heterogeneous resources, dynamic and Virtualization
• Heterogeneity is limited to a single organization	• Composed of heterogeneous resources from multiple organizations
• Static and has no concept of virtualization	• Virtualization and dynamic sharing is not limited to information, applications and hardware.
• Enable information sharing within a single organization	• Grids enable resource sharing among VOs (composed of multiple organizations)
• Resource discovery and monitoring on a global scale support is missing in distributed systems	• Resource discovery and monitoring on a global scale
• Less concerned with quality of service. Further they do not have a notion of trust as in grid systems	• They provide very specialized services and more concerned with quality of service have a notion of trust

1.9.2 Motivations for Using a Grid

In this section we discuss the advantages gained by using grids over conventional systems. Some of these motivations stem from the definition of grid in terms of VO. The others can be explained in terms of the grid as a high throughput computing system. It is important to have an understanding of these concepts, as they form the basis for the architecture of grids.

1.9.2.1 Enabling Formation of Virtual Organizations

- Grids enable collaboration among multiple organizations for sharing of resources.
- This collaboration is not limited to file exchange and implies direct access to computing resources.
- Members of the grid can dynamically be organized into multiple virtual organizations.
  - Each of these VOs may have different policies and administrative control.
  - All the VOs are part of a large grid and can share resources.
  - The resources shared among VOs may be data, special hardware, processing capability and information dissemination about other resources in the grid.
  - VOs hide the complexity of the grid from the user, enabling virtualization of heterogeneous grid resources.
- Members of a grid can be part of multiple VOs at the same time. Grids can be used to define security policies for the members enabling prioritization of resources for different users.

1.9.2.2 Fault Tolerance and Reliability

- The grid makes provision for automatic resubmission of jobs to other available resources when a failure is detected.
- Data grid can be defined as a grid for managing and sharing a large amount of distributed data.

Example 1: Suppose a user submits his job for execution at a particular node in the grid. The job allocates appropriate resources based on availability and the scheduling policy of the grid. Now suppose that the node, which is executing the job crashes due to some reason.

Example 2: Data grids serve multiple purposes. They can be used to increase the file transfer speed. Several copies of data can be created in geographically distributed areas. If a user needs the data for any computational purpose, it can be accessed from the nearest machine hosting the data. They increase overall computational efficiency. Further, if some of the machines in the data grid are down, other machines can provide the necessary backup. If it is known in advance that a particular machine will be accessing the data more frequently than others, data can be hosted on a machine near to that machine.

Both these examples illustrate the concept of virtualization. In the first example the user knows nothing about the grid failure. In the second example, the user accessing the data does not know which machine in the system serves his/her request.

1.9.2.3 Balancing and Sharing Varied Resources

Balancing and sharing resources are an important aspect of grids, which provide the necessary resource management features.

- This aspect enables the grid to evenly distribute the tasks to the available resources.
- Suppose a system in the grid is over-loaded. The grid scheduling algorithm can reschedule some of the tasks to other systems that are idle or less loaded.
- In this way the grid scheduling algorithm transparently transfers the tasks to a less loaded system thereby making use of the underutilized resources.

1.9.2.4 Parallel Processing

[1] Some tasks can be broken into multiple subtasks, each of which could be run on a different machine.

- Examples of such tasks can be mathematical modeling, image rendering or 3D animation. Such applications can be written to run as independent subtasks and then the results from each of these subtasks can be combined to produce the desired output.

[2] Constraints on dividing the task into subtask, using same data structures, types of tasks

- Limit on the number of subtasks which a task can be divided, limiting the maximum achievable performance increase.
- If two or more of these subtasks are operating on the same set of data structures, then some locking mechanism must exist so that the data structure does not become inconsistent.
- So there exists a constraint on the types of tasks, which can be made to run as a grid application and there also exists a limit to which an application can be made grid-enabled.

1.9.2.5 Quality of Service (QoS)

A grid can be used in a scenario where users submit their jobs and get the output, and then they are charged based on some metric like time taken to complete the task. In such scenarios the services delivered to the user expect certain quality of service.

Service Level Agreement (SLA)

SLA specifies the minimum quality of service, availability, etc, expected by the user and the charges levied on those services. SLA can specify the minimum expected up-time for the system.

Grid Scheduling Algorithm

- Based on the requirement of the user, his/her task could be given priority over other users' tasks by the grid scheduling algorithm.
  - ✓ For example, a user may require the services of the grid for a real-time application and thus has a more stringent QoS requirement than some other users. So, the grid scheduler could give his/her job more priority than other jobs and thus provide the necessary QoS to the user's real-time application.
- QoS can also be provided by reserving grid resources for certain jobs. If the resource reserved for a user's specific job is free for a while, it can report its status to a resource management node in the grid. The resource can then be used by the grid for its use until it is free.
  - ✓ For example, if it is a computing resource, it may be used by the grid for execution of other jobs in the grid. As soon as the requirement for the reserved resource arises, the jobs utilizing these resources are preempted and make way for the higher priority jobs (the job for which the resources were reserved). The preempted job is put in the job queue along with the information on its completion status. This job can be scheduled by the grid scheduler once there are available resources in the grid.

Grids are different because they provide such features on a multi institutional level and thus enable management of geographically distributed resources. Distributed systems that provide such features generally operate on an organizational level and have a centralized point of control unlike the grids.

1.4 Grid Architecture: Basic Concepts

Grid architecture can be visualized as a layered architecture figure 1.22

- The higher layers are focused on the user (user-centric)
- The lower layers are more focused on computers and networks (hardware-centric)

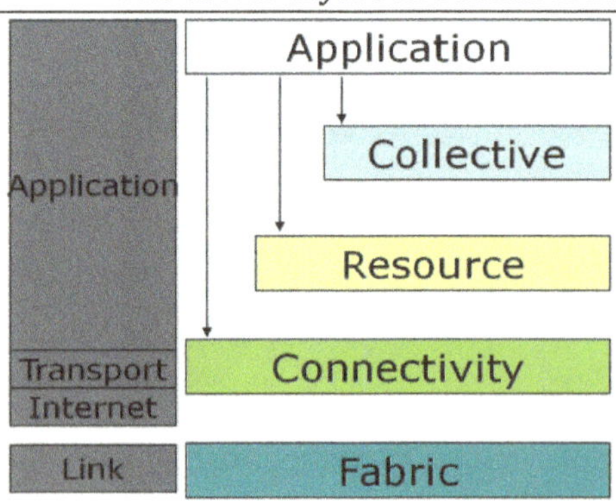

Coordinating multiple resources: ubiquitous infrastructure services, app-specific distributed services

Sharing single resources: negotiating access, controlling use

Talking to things: communication (Internet protocols) & security

Controlling things locally: Access to and control of, resources

Figure 1. 22 Layered Grid Architecture

- At the base of everything, the **bottom layer** is the **network**, which assures the connectivity for the resources in the Grid.
- On top of it lies the **resource layer**, made up of the actual resources that are part of the Grid, such as computers, storage systems, electronic data catalogues, and even sensors such as telescopes or other instruments, which can be connected directly to the network.
- The **middleware layer** provides the tools that enable the various elements (servers, storage, networks, etc.) to participate in a unified Grid environment. The middleware layer can be thought of as the intelligence that brings the various elements together - the "brain" of the Grid.
- The highest layer of the structure is the **application layer**, which includes all different user applications (science, engineering, and business, financial), portals and development toolkits supporting the applications. This is the layer that users of the grid will see.

In most common Grid architectures, the application layer also provides the so-called **serviceware**, the sort of general management functions such as

- measuring the amount a particular user employs the Grid
- billing for this use
- keeping accounts of who is providing resources and who is using them
- Keeps track of sharing the resources of a variety of institutions amongst large numbers of different users.
- The serviceware is in the top layer, because it is something the user interacts with, whereas the middleware is a "hidden" layer that the user should not have to worry about.

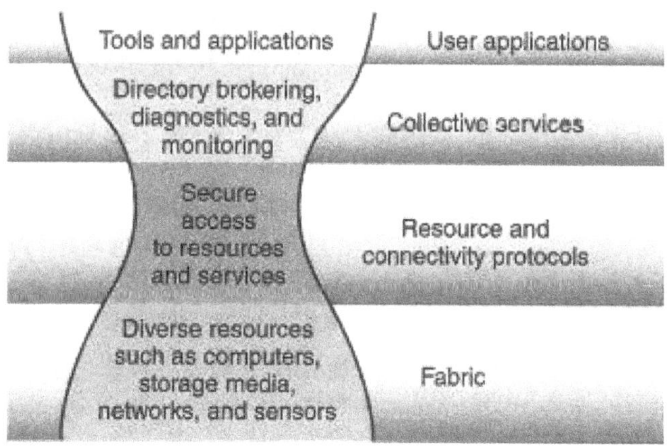

Figure 1. 23 Layered structure Grid Architecture

Fabric

- There are other ways to describe this layered structure. Figure 2.23

- The term **fabric** for all the physical infrastructure of the Grid, including computers and the communication network.
- Within the middleware layer, distinctions can be made between a layer of resource and connectivity protocols, and a higher layer of collective services.

Resource and connectivity Protocols

- **Resource and connectivity protocols:** handle all "Grid specific" network transactions between different computers and other resources on the Grid. Remember that the network used by the Grid is the Internet, the same network used by the Web and by many other services such as e mail.
- **Communication protocols**: A myriad of transactions is going on at any instant on the Internet, and computers that are actively contributing to the Grid have to be able to recognize those messages that are relevant to them, and filter out the rest. This is done with communication protocols, which let the resources speak to each other, enabling exchange of data.
- **Authentication protocols:** which provide secure mechanisms for verifying the identity of both users and resources

Collective Services

The collective services are also based on protocols: **information protocols**, which obtain information about the structure and state of the resources on the Grid, and **management protocols** which negotiate access to resources in a uniform way.

The services include:

- keeping directories of available resources updated at all times
- brokering resources (which like stock broking, is about negotiating between those who want to "buy" resources and those who want to "sell")
- monitoring and diagnosing problems on the Grid
- replicating key data so that multiple copies are available at different locations for ease of use
- providing membership/policy services for keeping track on the Grid of who is allowed to do what, when

User Applications

The topmost layer is the **applications layer**. Applications rely on all the other layers below them in order to run on the Grid. To take a fairly concrete example, consider a user application that needs to analyze data contained in several independent files. It will have

- obtain the necessary authentication credentials to open the files (resource and connectivity protocols)
- query an information system and replica catalogue to determine where copies of the files in question can currently be found on the Grid, as well as where computational
- resources to do the data analysis are most conveniently located (collective services)
- submit requests to the fabric - the appropriate computers, storage systems, and networks - to extract the data, initiate computations, and provide the results (resource and connectivity protocols)

- **monitor** the progress of the various computations and data transfers, notifying the user when the analysis is complete, and detecting and responding to failure conditions (collective services).

1.4.1 Security

Security forms the vital aspect of grid computing, three most desirable security features a grid should provide. These are

- **Single sign-on** means that the user is able to login once using his security credentials and can then access the service of the grid for a certain duration
- **Authentication** refers to providing the necessary proof to establish one's identity. So, when you login to your email account, you authenticate to the server by providing your username and password.
- **Authorization** is the process that checks the privileges assigned to a user. For example, a website may have two kinds of user, a guest user and a registered user.

A guest user may be allowed to perform basic tasks while the registered user may be allowed to perform a range of tasks based on his preferences. Authorization is performed after the identity of a user has been established through authentication. Other components of the grid that are part of security infrastructure are credential management and delegation of privileges.

1.4.2 Resource Management

A grid must optimize the resources under its disposal to achieve maximum possible throughput. Resource management includes submission of a job remotely, checking its status while it is in progress and obtaining the output when it has finished execution.

When a job is submitted,

- The available resources are discovered through a directory service.
- Then, the resources are selected to run the individual job.
- This decision is made by another resource management component of the grid, namely, the grid scheduler.
- The scheduling decision can be based on a number of factors.

For example, if an application consists of some jobs that need sequential execution because the result of one job is needed by another job, then the scheduler can schedule these jobs sequentially. The scheduling decision can also be based on the priority of the user's job as specified in the SLA.

1.4.3 Data Management

Data management in grids needed for managing large amounts of data.

This includes

- Secure data access
- Replication and migration of data
- Management of metadata
- Indexing
- Data-aware scheduling
- Caching etc.

Data aware-scheduling means that scheduling decisions should take into account the location of data. For example, the grid scheduler can assign a job to a resource located close to data instead of transferring large amounts of data over the network, which can have significant performance overheads. Suppose the job has been scheduled to run on a system that does not have the data needed for the job. This data must be transferred to the system where the job will execute. So, a grid data management module must provide a secure and reliable way to transfer data within the grid.

1.4.4 Information Discovery and Monitoring

The grid scheduler *needs to be aware of the available resources* to allocate resources for carrying out a job.

This information is obtained from an *information discovery service* running in the grid.

- It contains a list of resources available for the disposal of the grid and their current status.
- When a grid scheduler queries the information service for the available resources, it can put constraints such as finding those resources that are relevant and best suited for a job.
- By relevance of resource we mean those resources which can be used for the job.

The computing capacity *needed for a job and the job requires fast CPUs for its execution*, we select only those machines fast enough for the timely completion of the job.

The information discovery service can function in two ways.

- It can publish the status of available resources through a defined interface (web services) or it can be queried for the list of available resources.
- The information discovery service can be organized in a hierarchical fashion, where the lower information discovery services provide information to the one situated above it.
- The hierarchical structure brings about the flexibility needed for grids, which contains a vast amount of resources, because it can become practically impossible to store the information about all the available resources in one place.

1.5 Some Standards for Grid

In this section we look at some of the open standards used for implementing a grid.

1.5.1 Web Services

Grid services, defined by OGSA, are an extension of web services. So, grid service can leverage the available web services specifications. Here we discuss the most basic web service standards. The four basic web service specifications are:

[1] **EXtensible Markup Language (XML)** - XML is a markup language whose purpose is to facilitate sharing of data across different interfaces using a common format. It forms the basis of web services. All the messages exchanged in web services adhere to the XML document format.

[2] **Simple Object Access Protocol (SOAP)** - SOAP is a message-based communication protocol, which can be used by two parties communicating over the Internet. SOAP messages are based on XML and are hence platform independent. It forms the

foundation of the web services protocol stack. SOAP messages are transmitted over HTTP. So unlike other technologies like RPC or CORBA, SOAP messages can traverse a firewall. SOAP messages are suitable when small messages are sent. When the size of message increases, the overhead associated with it also increases and hence the efficiency of the communication decreases.

[3] **Web Service Definition Language (WSDL)** - WSDL is an XML document used to describe the web service interface.

A WSDL document describes a web service using the following major elements:

a) Port Type - The set of operations performed by the web service. Each operation is defined by a set of input and output messages.
b) Message - It represents the messages used by the web service. It is an abstraction of the data being transmitted.
c) Types - It refers to the data types defined to describe the message exchange.
d) Binding - It specifies the communication protocol used by the web service.
e) Port - It defines the binding address for the web service.
f) Service - It is used for aggregating a set of related ports

[4] **Universal Description, Discovery and Integration (UDDI)** - UDDI is an XML-based registry used for finding a web service on the Internet. It is a specification that allows a business to publish information about it and its web services allowing other web services to locate this information. A UDDI registry is an XML-based service listing. Each listing contains the necessary information required to find and bind to a particular web service.

1.5.2 Open Grid Services Architecture (OGSA)

Open Grid Services Architecture (OGSA) defines

- a web services based framework for the implementation of a grid.
- It seeks to standardize service provided by a grid such as resource discovery, resource management, security, etc, through a standard web service interface.
- It also defines the features that are not necessarily needed for the implementation of a grid.

OGSA is based on existing web services specifications and adds features to web services to make it suitable for the grid environment.

1.5.3 Open Grid Services Infrastructure (OGSI)

OGSA describes the features that are needed for the implementation of services provided by the grid, as web services.

- Provides a formal and technical specification needed for the implementation of grid services.
- It provides a description of Web Service Description Language (WSDL), which defines a grid service.
- Provides the mechanisms for creation, management and interaction among grid services.

1.5.4 Web Services Resource Framework (WSRF)

The motivation behind development of WS-Resource Framework is to define a "generic and open framework for modeling and accessing stateful resources using web services"

- It defines conventions for state management enabling applications to discover and interact with stateful web services in a standard way.
- Grid-based applications need the notion of state because they often perform a series of requests where output from one operation may depend on the result of previous operations.
- Used to develop such stateful grid services.
- The format of message exchange is defined by the WSDL.

1.5.5 OGSA-DAI

Open Grid Services Architecture-Data Access and Integration (OGSA-DAI) aim is to develop middleware to provide access and integration to distributed data sources using a grid.

- This middleware provides support for various data sources such as relational and XML databases.
- These data sources can be queried, updated and transformed via OGSA-DAI web service.
- These web services can be deployed within a grid, thus making the data sources grid enabled.
- The request to OGSA-DAI web service to access a data source is independent of the data source served by the web service.
- OGSA web services are compliant with Web Services Inter-operability (WS-I) and WSRF specifications, the two most important specifications for web services

2. GRID SERVICES

2.1 Grid Architecture and Service Modeling

The grid is a metacomputing infrastructure that brings together computers (PCs, workstations, server clusters, supercomputers, laptops, notebooks, mobile computers, PDAs, etc.) to form a large collection of compute, storage, and network resources to solve large-scale computation problems or to enable fast information retrieval by registered users or user groups. The coupling between hardware and software with special user applications is achieved by leasing the hardware, software, middleware, databases, instruments, and networks as computing utilities. Good examples include the renting of expensive special-purpose application software on demand and transparent access to human genome databases.

The goal of grid computing is to explore fast solutions for large-scale computing problems. Grid computing takes advantage of the existing computing resources scattered in a nation or internationally around the globe. In grids, resources owned by different organizations are aggregated together and shared by many users in collective applications.

Grids rely heavy use of LAN/WAN resources across enterprises, organizations, and governments. The virtual organizations or virtual supercomputers are new concept derived from grid or cloud computing. These are virtual resources dynamically configured and are not under the full control of any single user or local administrator.

2.1.1 Grid History and Service Families

Network-based distributed computing becomes more and more popular among the Internet users. Ever since the 1990s, grids became gradually available to establish large pools of shared resources. The approach is to link many Internet applications across machine platforms directly in order to eliminate isolated resource islands.

The idea of the grid was pioneered by Ian Foster, Carl Kesselman and Steve Tuecke in a 2001 paper. With is ground work, they are often recognized as the fathers of the grids. The Globus Project supported by DARPA has promoted the maturity of grid technology with a rich collection of software and middleware tools for grid computing. In 2007, the concept of cloud computing was thrown out, which in many ways was extending grid computing through virtualized data centers. In this beginning section, we introduce major grid families and review the grid service evolution over the past 15 years.

Grids differ from conventional HPC clusters. Cluster nodes are more homogeneous machines that are better coordinated to work collectively and cooperatively. The grid nodes are heterogeneous computers that are more loosely coupled together over geographically dispersed sites. In 2001, Forbes Magazine advocated the emergence of the great global grid (GGG) as a new global infrastructure. This GGG evolved from the World Wide Web (WWW) technology we have enjoyed for many years. Four major families of grid computing systems were suggested by the Forbes GGG categorization as summarized in Table 2.1.

Table 2.1 Four Grid Families Identified in the Great Global Grid (GGG)

Grid Family	Representative Grid Systems and References
Computational Grids or Data Grids	TeraGrid (US), EGEE (EU), DataGrid (EU), Grid'5000 (france), ChinaGrid (China), NAS (NASA), LCG (Cern), e-Science (UK), D-Grid (Nordic), FutureGrid (US), etc.
Information Grids or Knowledge Grids	Semantic Grid, Ontology Platform, BOINC (Berkeley), D4Science, Einsten@Home, Information Power Grid (NASA)
Business Grids	BEinGrid (EU), HP eSpeak, IBM WebSphere, Sun Grid Engine, Microsoft .NET, etc.
P2P/Volunteer Grids	SETI@Home, Parasic Grid, FightAIDS@Home, Foldong@Home, GIMPS, etc.

2.1.1.1 Four Grid Service Families

Most of today's grid systems are called computational grids or data grids. Good examples are the NSF TeraGrid installed in the United States and the DataGrid built in the European Union. Information or knowledge grids post another grid class dedicated to knowledge management and distributed ontology processing. The Semantic web, also known as semantic grids, belongs to this faimly.

Ontology platform falls into information or knowledge grids. Other information/knowledge grids include the Berkeley BOINC and NASA's Information Power Grid. In the business world, we see a family, called business grids, built for business data/information processing. These are represented by the HP eSpeak, IBM WebSphere, Microsoft .NET, and Sun One systems. Some business grids are being transformed into Internet clouds. The last grid class includes several grid extensions such as P2P grids and parasitic grids.

2.1.1.2 Grid Resources

Table 2.2 summarizes typical resources that are required to perform grid computing. Many existing protocols (IP, TCP, HTTP, FTP, and DNS) or some new communication protocols can be used to route and transfer data. The resource layer is responsible for sharing single resources. An interface is needed to claim the static structure and dynamic status of local resources.

Table 2.2 Control Operations and Enquiries for Aggregating Grid Resources

Resources	Control Operations	Enquiries
Compute resources	Starting, monitoring, and controlling the execution of resultant processes; control over resources: advance reservation	Hardware and software characteristics; relevant load information: current load and queue state
Storage resources	Putting and getting files; control over resources allocated to data transfers: advance reservation	Hardware and software characteristics; relevant load information: available space and bandwidth utilization

Network resources	Control over resources allocated	Network characteristics and load
Code repositories	Managing versioned source and object code	Software files and compile support
Service catalogs	Implementing catalog query and update operations: a relational database	Service order information and agreements The

The grid should be able to accept resource requests, negotiate the Quality of Service (QoS), and perform the operations specified in user applications. The collective layer handles the interactions among a collection of resources. This layer implements functions such as resource discovery, co-allocation, scheduling, brokering, monitoring, and diagnostics. Other desired features include replication, grid-enabled programming, workload management, collaboration, software discovery, access authorization, and community accounting and payment. The application layer comprises mainly user applications. The applications interact with components in other layers by using well-defined APIs (application programming interfaces) and SDKs (software development kits).

2.2 Open Grid Services Architecture (OGSA)

The OGSA is an open source grid service standard jointly developed by academia and the IT industry under coordination of a working group in the Global Grid Forum (GGF).

- The standard was specifically developed for the emerging grid and cloud service communities. The OGSA is extended from web service concepts and technologies.
- The standard defines a common framework that allows businesses to build grid platforms across enterprises and business partners.
- The intent is to define the standards required for both open source and commercial software to support a global grid infrastructure.

2.2.1 OGSA Framework

- The OGSA was built on two basic software technologies:
  - ➔ Globus Toolkit widely adopted as a grid technology solution for scientific and technical computing, and
  - ➔ Web services (WS 2.0) as a popular standards-based framework for business and network applications. The
- OGSA is intended to support the creation, termination, management, and invocation of stateful, transient grid services via standard interfaces and conventions.
- The OGSA framework specifies the physical environment, security, infrastructure profile, resource provisioning, virtual domains, and execution environment for various grid services and API access tools.
- Compared with the layered grid architecture, the OGSA is service-oriented. A service is an entity that provides some capability to its client by exchanging messages.
- The service-oriented architecture (SOA) serves as the foundation of grid computing services.
- The individual and collective states of resources are specified in this service standard.

The standard also specifies interactions between these services within the particular SOA for grids.

- The architecture is not layered, where the implementation of one service is built upon modules that are logically dependent. One may classify this framework as object-oriented. Many web service standards, semantics, and extensions are applied or modified in the OGSA.

2.2.2 OGSA Interfaces

The OGSA is centered on grid services. These services demand special well-defined application interfaces. These interfaces provide

- resource discovery
- dynamic service creation
- lifetime management
- notification
- manageability

The conventions must address naming and upgradeability. Table 2.3 summarizes the interfaces proposed by the OGSA working group.

Port Type	Operation	Brief Description
Grid service	Find service data	Query a grid service instance, including the handle, reference, primary key, home handle map, interface information, and service-specific information. Extensible support for various query languages.
	Termination time	Set (and get) termination time for grid service instance.
	Destroy	Terminate grid service instance.
Notification source	Subscribe to notification topic	Subscribe to notifications of service events. Allow delivery via third-party messaging services.
Notification sink	Deliver notification	Carry out asynchronous delivery of notification messages.
Registry	Register service	Conduct soft-state registration of Grid Service Handles (GSHs).
	Unregister service	Unregister a GSH.
Factory	Create service	Create a new grid service instance.
Handle map	Find by handle	Return the Grid Service Reference (GSR) associated with the GSH.

Table 2.3 OGSA Grid Service Interfaces Developed by the OGSA Working Group

Two key properties of a grid service are

- **transience** - Being transient means the service can be created and destroyed dynamically
- **statefulness** - refers to the fact that one can distinguish one service instance from another

These properties have significant implications regarding how a grid service is named, discovered, and managed.

2.2.3 Grid Service Handle

- A GSH is a globally unique name that distinguishes a specific grid service instance from all others.

- The status of a grid service instance could be that it exists now or that it will exist in the future.
- These instances carry **no protocol or instance-specific addresses or supported protocol bindings**. Instead, these information items are encapsulated along with all other **instance-specific information.**
  - In order to interact with a specific service instance, a single abstraction is defined as a GSR, an the instance can change over the lifetime of the service.
  - The OGSA employs a **"handle-resolution"** mechanism for mapping from a GSH to a GSR.
  - The GSH must be globally defined for a particular instance. However, the GSH may not always refer to the same network address. A service instance may be implemented in its own way, as long as it obeys the associated semantics.

2.2.4 Grid Service Migration

- This is a mechanism for creating new services and specifying assertions regarding the lifetime of a service.
- The OGSA model defines a standard interface, known as a factor, to implement this reference.
- Any service that is created must address the former services as the reference of later services.
- The factory interface is labeled as a Create Service operation in Table 2.3.
- This creates a requested grid service with a specified interface and returns the GSH and initial GSR for the new service instance.
- It should also register the new service instance with a handle resolution service. Each dynamically created grid service instance is associated with a specified lifetime.

Example: Grid Service Migration Using GSH and GSR

Figure 2.1 shows how a service instance may migrate from one location to another during execution. A GSH resolves to a different GSR for a migrated service instance before (on the left) and after (on the right) the migration at time T. The handle resolver simply returns different GSRs before and after the migration. The initial lifetime can be extended by a specified time period by explicitly requesting the client or another grid service acting on the client's behalf.

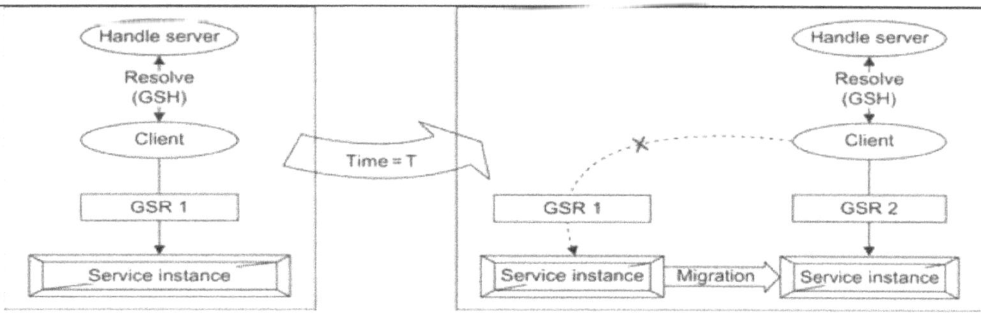

Figure 2.1 A GSH resolving to a different GSR for a migrated service instance before (shown on the left) and after (on the right) the migration at time T.

If the time period expires without having received a reaffirmed interest from a client, the service instance can be terminated on its own and release the associated resources accordingly. The lifetime management enables robust termination and failure detection. This is done by clearly defining the lifetime semantics of a service instance. Similarly, a hosting environment is guaranteed to consume bounded resources under some system failures. If the termination time of a service is reached, the hosting environment can reclaim all resources allocated.

2.2.5 OGSA Security Models

- The OGSA supports security enforcement at various levels, as shown in Figure 2.2.

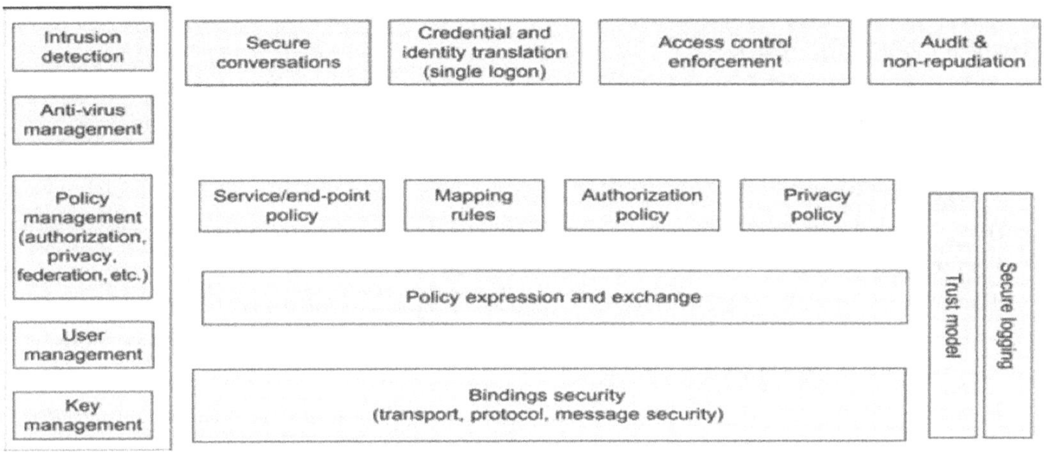

Figure 2.2 The OGSA security model implemented at various protection levels

- The grid works in a **heterogeneous distributed environment**, which is essentially open to the general public.
- It must be able to **detect intrusions or stop viruses** from spreading by implementing secure conversations, single logon, access control, and auditing for nonrepudiation.
- At the **security policy and user levels** want to apply a service or endpoint policy, resource mapping rules, authorized access of critical resources, and privacy protection.
- At the **Public Key Infrastructure (PKI) service level**, the OGSA demands security binding with the security protocol stack and bridging of certificate authorities (CAs), use of multiple trusted intermediaries, and so on. Trust models and secure logging are often practiced in grid platforms.

2.2.6 Detailed view of OGSA/OGSI

2.2.6.1. Key Aspects

There are two main logical components of OGSA:

i. The Web-services-plus-OGSI layer, and
ii. The OGSA-architected services layer.

Four main layers comprise the OGSA architecture

Grid applications layer

- This layer is the user-visible layer.
- It supports user applications.
- Eventually, a "rich" set of grid-architected services is expected to be developed.

OGSA-architected grid services layer

- Services in this layer include:
  - ➔ Discovery
  - ➔ Lifecycle
  - ➔ State management
  - ➔ Service Groups
  - ➔ Factory
  - ➔ Notification, and
  - ➔ Handle Map.

These services are based on the Web services layer.

- The GGF was working at press time to define many of these architected grid services in areas such as program execution, data services, and core services.

Web Services layer, plus the OGSI extensions that define grid services

- The OGSI specification defines grid services and builds on standard Web services technology.
- OGSI exploits the mechanisms of Web services such as XML and WSDL to specify standard interfaces, behaviors, and interaction for all grid resources.
- OGSI extends the definition of Web services to provide capabilities for dynamic, stateful, and manageable Web services that are required to model the resources of the grid.

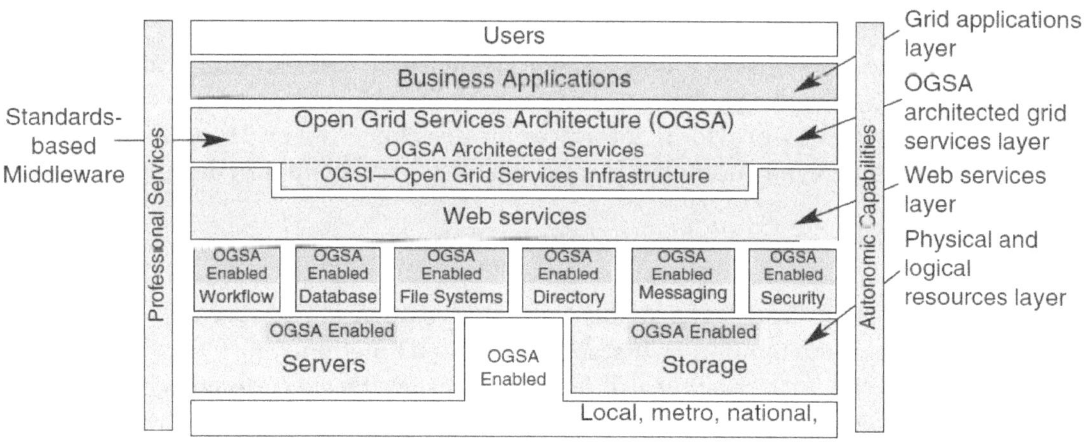

Figure 2.3 Basic functional model for grid environment

Physical and logical resources layer

- The concept of resources is central to OGSA and to grid computing in general.

- Resources comprise the capabilities of the grid.
- Physical resources include servers, storage, and network. Above the physical resources are logical resources.
- Logical resources provide additional function by virtualizing and aggregating the resources in the physical layer. General-purpose middleware such as file systems, database managers, directories, and workflow managers provide these abstract services on top of the physical grid.

The GGF OGSA working group found it necessary to augment core Web services functionality to address grid services requirements. OGSI extends Web services by introducing **interfaces and conventions in two main areas**

1. "Interfaces."

The dynamic and potentially transient nature of services in a grid:

- Particular service instances may come and go as work is dispatched, as resources are configured and provisioned, and as system state changes.
- Therefore, grid services need interfaces to manage the creation, destruction, and life-cycle management of these dynamic services.

2. "State."

Grid services typically have attributes and data associated with them. This is similar in concept to the traditional structure of objects in object-oriented programming:

- Objects have behavior and data. Likewise, Web services were found to be in need of being extended to support state data associated with grid services.
- Basic Web services are stateless (e.g., add, subtract).
- Most real-world applications involve stateful transactions [e.g., query (sd2), getdata (row3-row17)].
- State is linked to a "handle" or sessionID as a parameter.
- Protocols such as SOAP, SMTP, and FTP use state mechanisms (sessionID, packet headers, TCP sockets, respectively).

Consistent with these two observations, OGSI introduces an **interaction model for grid services**. The interaction model provides a uniform way for software developers to model and interact with grid services by providing interfaces for discovery, life cycle, state management, creation and destruction, event notification, and reference management (these services were depicted in Figure 2.4)

Below, we list interfaces and conventions that OGSI introduces

Factory

- A mechanism (interface) that provides a way to create new grid services.
- Factories may create temporary instances of limited function,
  - such as a scheduler creating a service to represent the execution of a particular job;
  - or they may create longer-lived services such as a local replica of a frequently used data set.
- Not all grid services are created dynamically;
  - for example, some services might be created as the result of an instance of a physical resource in the grid, such as a processor, storage, or network device.

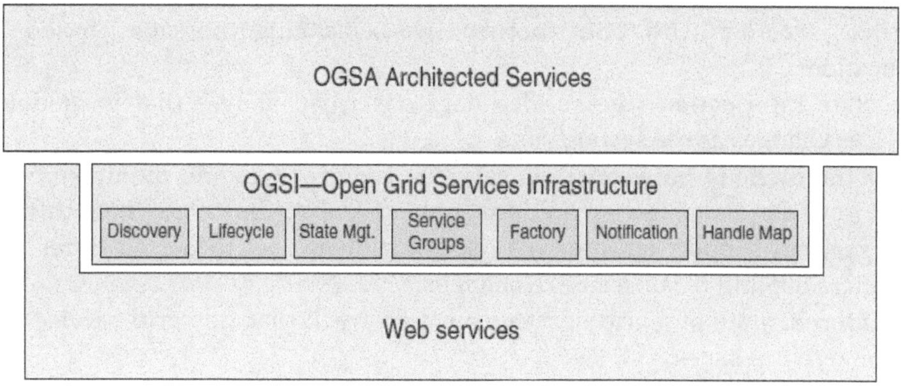

Figure 2.4 OGSA reliance on OGSI

Life cycle

- A mechanism architected to prevent grid services from consuming resources indefinitely without requiring a large-scale distributed "garbage collection" scavenger.
- Every grid service has a termination time set by the service creator or factory. Because grid services may be transient, grid service instances are created with a specified lifetime.
- The lifetime of any particular service instance can be negotiated and extended, as required, by components that are dependent on or manage that service.
    - In turn, a client with appropriate authorization can use termination time information to check the availability (lease period) of the service; the client can also request to extend the current lease time by sending a keep-alive message to the service with a new termination time.
    - If the service accepts this request, the lease time can be extended to the new termination time requested by the client.
- This soft-state life cycle is controlled by appropriate security and policy decisions of the service, and the service has the authority to control this behavior
    - for example, a service can arbitrarily terminate a service or can extend its termination time even while the client holds a service reference

State management

- OGSI specifies a framework for representing this state, called service data, and a mechanism for inspecting or modifying that state, named Find/SetServiceData.
- OGSI requires a minimal amount of state in service data elements that every grid service must support, and requires that all services implement the Find/SetServiceData portType.

Service groups

- Service groups are collections of grid services that are indexed (using service data described above) for some specific purpose.
    - For example, they might be used to collect all the services that represent the resources in a particular cluster node within the grid.

Notification

- Services interact with one another by exchanging messages based on service invocation.
- The state information (the service data described above) that is modeled for grid services changes as the system runs.
- Many interactions between grid services require dynamic monitoring of changing state. Notification applies a traditional publish/subscribe paradigm to this monitoring.
- Grid services support an interface (NotificationSource) to permit other grid services (NotificationSink) to subscribe to changes.
- The internal state of a grid service can keep track that this grid service has received one or zero messages.
- This reliable message delivery mechanism guaranteed by the internal state can build business-oriented transactions.
- In a transient stateful service, OGSA provides a mechanism to capture the state information associated with any operation that fails. If an operation fails, the keep-alive messages cease if there is no service client for invoking this running service instance.
- Then the grid service instance automatically times out and frees the computing resources associated with this service instance

Handle Map

- This deals with service identity.
- When Factories are used to create a new instance of a Grid Service, the Factory returns the identity of the newly instantiated service.
- This identity is composed of two parts: a Grid Service
    - Handle (GSH) and
    - a Grid Service Reference (GSR).
- A GSH provides a reference the grid service indefinitely; GSR can change within the grid services lifetime.

The **Handle Map interface** provides a way to obtain a GSR given a GSH.

- The user application invokes create Grid Service requests on the Factory interface to create a new service instance.
- The newly created service instance associated with the grid service interface will be automatically allocated computing resources.
- Meanwhile, an initial lifetime of the instance can be specified before the service instance is created. The newly created service instance will keep the user credentials for performing further interactions with other systems over the Internet.

The newly created grid service instance will be automatically assigned a globally unique name called the GSH, which is used to distinguish this specific service instance from other grid service instances

These enhancements are specified in OGSI. As the OGSI specification was finalized and implementations began to appear, some standards organizations became interested in incorporating a portion of the functionality outlined in OGSI within appropriate Web services standards; hence, over time, it is expected that much of the OGSI functionality will be incorporated in Web services standards.

2.2.6.2 Ancillary Aspects

Drilling down an additional level of detail, one can further categorize grid-architected services into four categories, as shown in Figure 2.5:

- Grid core services
- Grid program execution services
- Grid data services
- Domain-specific services

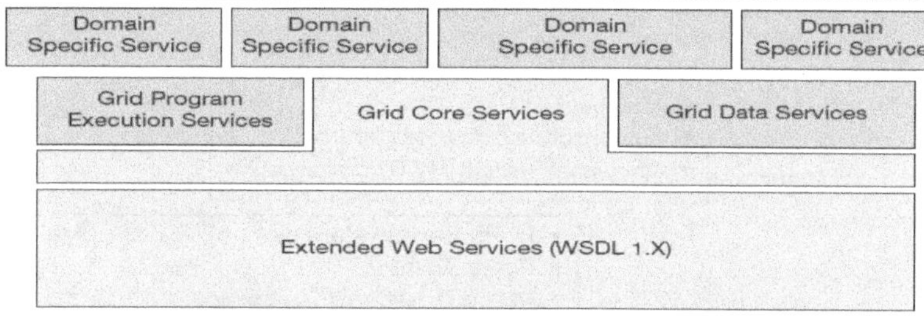

Figure 2.5 The structure of OGSA architected services

Grid Core Services

Figure 2.5 shows that the grid core services are composed of four main types of services:

1. Service management
2. Service communication
3. Policy management
4. Security

Unlike the OGSI functions that are largely implemented as extensions to basic Web Services protocols and an interaction model, these core services are actually implemented as grid services (upon the OGSI base). These services are considered core primarily because it is expected that they will be broadly exploited by most higher level services implemented either in support of program execution or data access, or as domain-specific services.

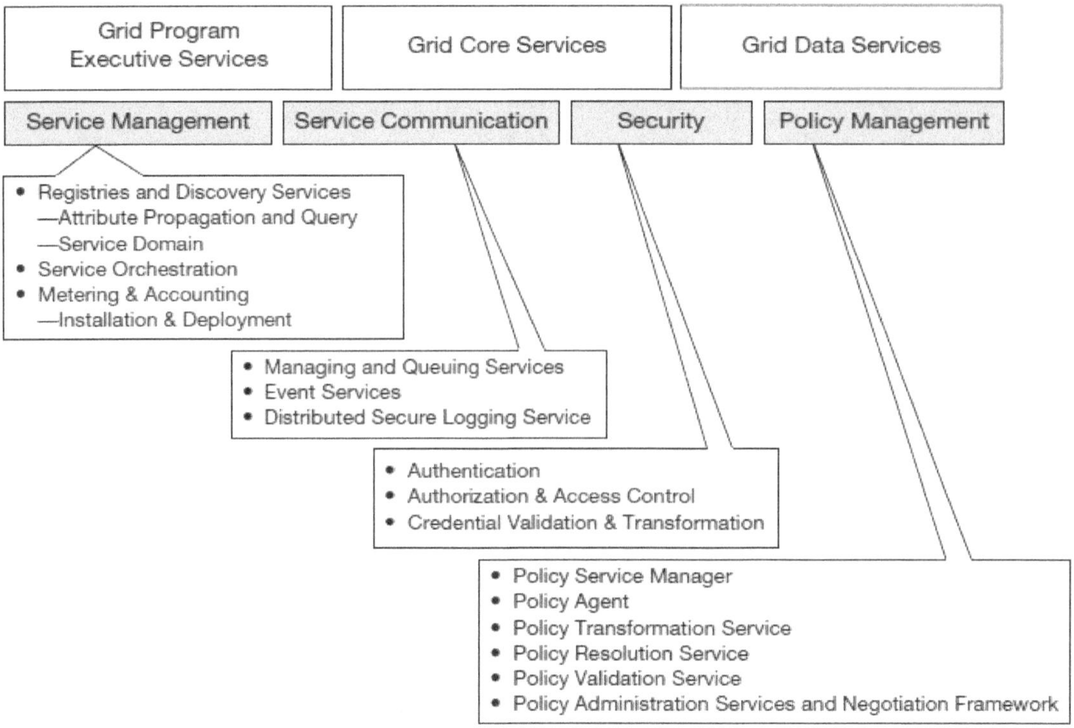

Figure 2.5 Grid core services

- **Service management**.
  - Service management provides functions that manage the services deployed in the distributed grid. It automates a variety of installation, maintenance, monitoring, and troubleshooting tasks within a grid system.
  - Service management includes functions
    - √ for provisioning and deploying the system components;
    - √ Includes functions for collecting and exchanging data about the operation of the grid.
    - √ This data is used for both "online" and "offline" management operations,
    - √ includes information about faults, events, problem determination, auditing, metering, accounting, and billing

- **Service communication**.
  - This includes a range of functions that support the basic methods for grid services to communicate with each other.
  - These functions support several communication models that may be composed to enable effective interservice communication, including queued messages, publish–subscribe event notification, and reliable distributed logging.
  - grid services can be published to a UDDI registry, or WSIL documents

- → the UDDI registry becomes a central place to store such information about and locations for grid services that enables publishing and searching of trading partners' businesses and their grid services.
- → There are two types of UDDI registries: private and public.
    - √ Application developers and/or service providers can publish the grid services to the public UDDI registries operated by IBM, Microsoft, HP, or SAP.
    - √ If one wants to publish one's own private or confidential grid services, one can use a private UDDI registry.

- o As an alternative, for testing purposes or for small-scale integration, a developer can publish the company's grid services to WSIL documents, since WSIL enables grid services discovery, deployment, and invocation without the need for a UDDI registry.
    - → WSIL provides the means for aggregating references of preexisting service description documents that have been authored in any number of formats; these inspection documents are then made available on a Web site.
- o Figure 2.6 illustrates an example grid service deployment and publishing diagram. The Remote Procedure Call servlet of SOAP and the real implementation of the grid services can be deployed on an application server. All the invocation messages will be captured by the SOAP Remote Procedure Call servlet that routes the messages to the corresponding grid service.

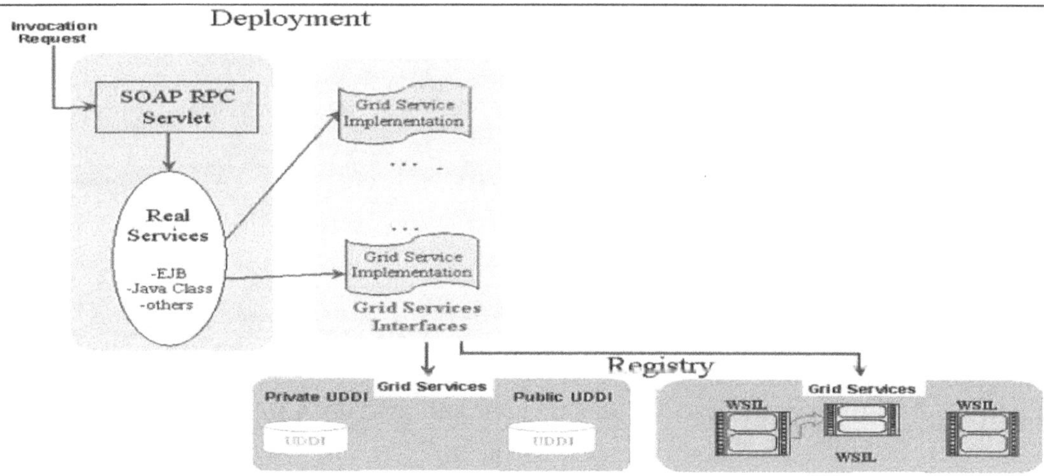

Figure 2.6 An example grid service deployment and publishing diagram.

- **Policy services**
    - o These create a general framework for creation, administration, and management of policies and agreements for system operation.
    - o Policy services include policies governing security, resource allocation, and performance, as well as an infrastructure for "policy-aware" services to use policies to govern their operation.

- Policy and agreement documents provide a mechanism for the representation and negotiation of terms between service providers and their clients (either user requests or other services); terms include specifications, requirements, and objectives for function, performance, and quality that the suppliers and consumers exchange and that they can then use to influence their interactions.

- **Security services**
    - Security services support, integrate, and unify popular security models, mechanisms, protocols, and technologies in a way that enables a variety of systems to interoperate securely.
    - These security services enable and extend core Web services security protocols and bindings and provide service-oriented mechanisms for authentication, authorization, trust policy enforcement, credential transformation.
- **Grid Program Execution Services**
    - Grid program execution services are depicted in Figure 2.7. Mechanisms for job scheduling and workload management implemented as part of this class of services are central to grid computing and the ability to virtualize processing resources.
    - Although OGSI and core grid services are generally applicable to any distributed computing system, the grid program execution service class is unique to the grid model of distributed task execution that supports high-performance computing, parallelism, and distributed collaboration.

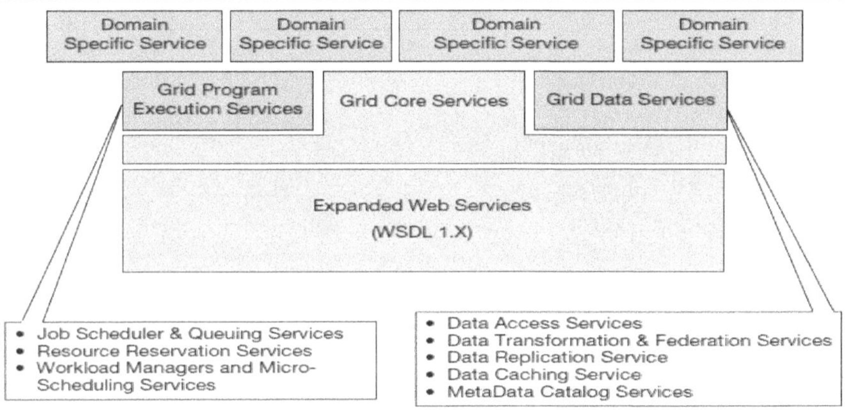

Figure 2.7 Grid program execution services and grid data services.

- **Grid Data Services**
    - Grid data services are also depicted in Figure 2.7. These interfaces support the concept of data virtualization and provide mechanisms related to distributed access to information of many types including databases, files, documents, content stores, and application-generated streams.
    - Services that comprise the grid data services class complement the computing virtualization conventions specified by program execution services (OGSA placing data resources on an equivalent level with computing resources).

  - Grid data services will exploit and vir tualize data using placement methods like data replication, caching, and high-performance data movement to give applications required QoS access across the distributed grid. Methods for federating multiple disparate, distributed data sources may also provide integration of data stored under differing schemas such as files and relational databases.

- **Domain-Specific Services**
  - The three categories discussed above (grid core services, grid program execution services, and grid data services) represent areas of active work by GGF research or working groups. Over time, as these services mature, domain-specific services can also be specified.
  - Domain-specific services will make use of the functionality that these services supply. It is critical that the GGF working groups are concentrating on specifying a broad set of useful grid services that software vendors and developers can then begin to implement.

2.2.6.3 Implementations of OGSI

As the core of the grid service architecture, OGSI needs to be hosted on a delivery platform that supports Web services. Vendors that offer OGSI implementations will likely directly use existing open source implementations provided by organizations like Globus, and/or they will integrate implementations with their hosting platform. However, grid architected services provide some opportunities for vendors and organizations to compete and differentiate themselves. This competition will create an "economy" of grid software providers whose innovation will help drive the acceptance of standards like OGSI/OGSA, and this will allow customers to build systems out of interoperable components. Areas of functionality in grid program execution and data services will require innovation and novel approaches, and these may well speed the market acceptance of grid solutions and provide market opportunities for vendors.

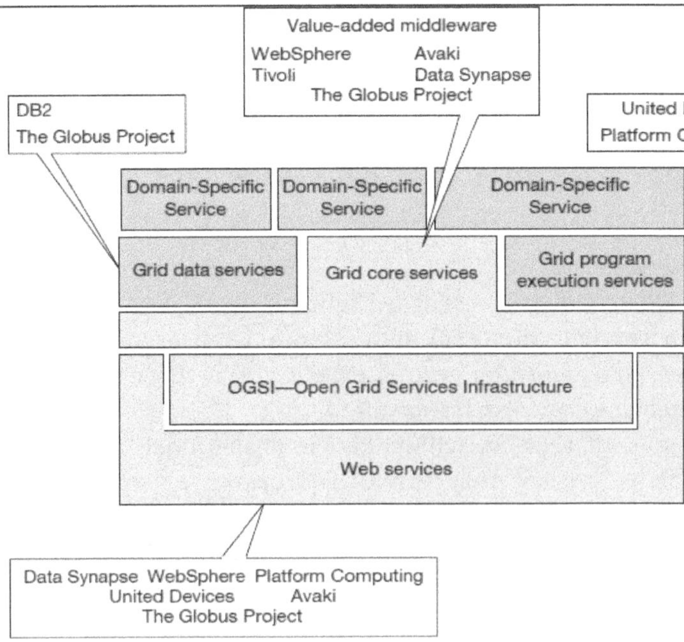

Figure 2.8 Grid program execution and data services hosting.

In Figure 2.8, that grid core services are likely to see a mix of open source reference implementations and vendor-provided "value added" implementations. OGSI/OGSA extends Web Services, it is not clear that software vendors will be able to differentiate themselves based on the quality of their core services implementations; such differention will likely be based on the business creativity and/or import of their domain-specific implementations.

For OGSA to grow in acceptance it needs to be implemented on multiple hosting platforms.

- The Globus Toolkit 3 (GT3) historically was the first full-scale implementation of the OGSI standard.
- GT3 was developed by the Globus Project, a research and development project focused on enabling the application of grid concepts to scientific, engineering, and commercial computing.
- It is expected that many of the OGSI implementations will be delivered via the open source development model and that existing reference implementations (GT3) will be used unmodified in appropriate hosting environments.
- GT3 is written in Java language using the J2EE framework; however, nothing limits OGSI from being implemented in other programming languages and hosted in other environments (the term "hosting environment" is used to denote the server in which one or more grid service implementations run).
- Figure 2.9 shows that a Java implementation of OGSI can be hosted on any of several J2EE environments (such as JBOSS, WebSphere, or BEA Weblogic).
- However, alternative platforms such as a traditional C or C++ environment or C# and Microsoft .NET are other possible hosting environments.
- Ideally, a small number of core implementations of OGSI (perhaps one per hosting platform) will be jointly developed by the industry and used in many products

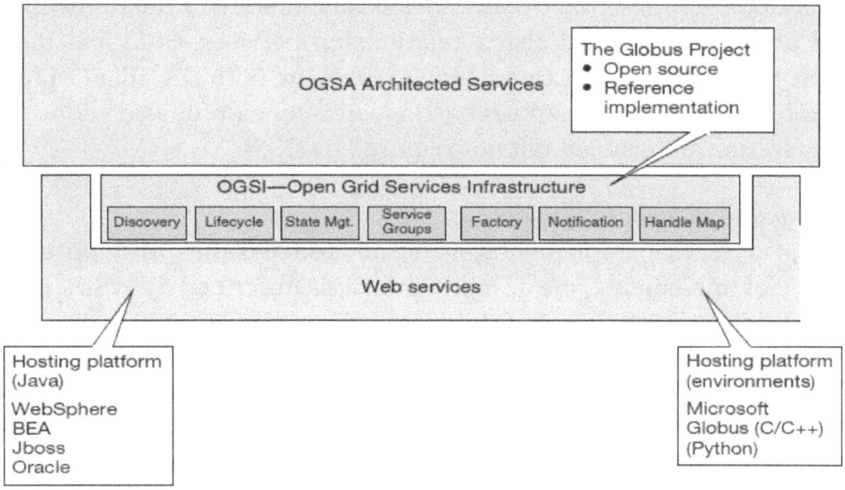

Figure 2.9 OGSI and web services hosting.

2.2.7 A More Detailed View of OGSA/OGSI

- The OGSA integrates key grid technologies including the Globus Toolkit with Web services mechanisms to create a distributed system framework based on the OGSI.
- A grid service instance is a (potentially transient) service that conforms to a set of conventions, expressed as WSDL interfaces, extensions, and behaviors, for such purposes as lifetime management, discovery of characteristics, and notification.
- OGSI introduces standard factory and registration interfaces for creating and discovering grid services.
- OGSI defines a component model that extends WSDL and XML schema definition to incorporate the concepts of
    - ➔ Stateful Web services
    - ➔ Extension of Web services interfaces
    - ➔ Asynchronous notification of state change
    - ➔ References to instances of services
    - ➔ Collections of service instances
    - ➔ Service state data that augment the constraint capabilities of XML schema definition
- The OGSI V1.0 specification proposes detailed specifications for the conventions that govern how clients create, discover, and interact with a grid service instance. That is, it specifies
    (1) How grid service instances are named and referenced;
    (2) The base, common interfaces (and associated behaviors) that all grid services implement;
    (3) The additional (optional) interfaces and behaviors associated with factories and service groups.

2.2.7.1 Setting the Context

GGF calls OGSI the "base for OGSA." Specifically, there is a relationship between OGSI and distributed object systems and also a relationship between OGSI and the existing (and evolving) Web services framework. One needs to examine both the client-side programming patterns for grid services and a conceptual hosting environment for grid services. The patterns described in this section are enabled but not required by OGSI.

2.2.7.1.1 Relationship to Distributed Object Systems

➔ A given grid service implementation is an **addressable and potentially stateful instance** that implements one or more interfaces described by WSDL portTypes.
- Grid service factories used to create instances implementing a given set of portType(s).
- Each grid service instance has a notion of identity with respect to the other instances in the distributed grid.
- Each instance can be characterized as state coupled with behavior published through type-specific operations.
- The architecture also supports introspection in that a client application can ask a grid service instance to return information describing itself, such as the collection of portTypes that it implements.

➔ Grid service instances are made accessible to (potentially remote) client applications through the use of a grid service handle and a grid service reference (GSR).
- These constructs are basically network-wide pointers to specific grid service instances hosted in (potentially remote) execution environments.
- A client application can use a grid service reference to send requests, represented by the operations defined in the portType(s) of the target service description directly to the specific instance at the specified network-attached service endpoint identified by the grid service reference.

➔ Client stubs and helper classes isolate application programmers from the details of using grid service references. Some client-side infrastructure software assumes responsibility for directing an operation to a specific instance that the GSR identifies.

➔ OGSI does not adopt the term distributed object model or distributed object system when describing stateful instances, typed interfaces, global names, etc concepts, but instead uses the term "open grid services infrastructure," thus emphasizing the connections that are established with both Web services and grid technologies.

2.2.7.1.2 Client-Side Programming Patterns

Another important issue is how OGSI interfaces are likely to be invoked from client applications. OGSI exploits an important component of the Web services framework:

- The use of WSDL to describe multiple protocol bindings, encoding styles, messaging styles (RPC versus document oriented), and so on, for a given Web service.
- The Web Services Invocation Framework (WSIF) and Java API for XML RPC (JAX-RPC) are among the many examples of infrastructure software that provide this capability.

Client-side architecture for OGSI

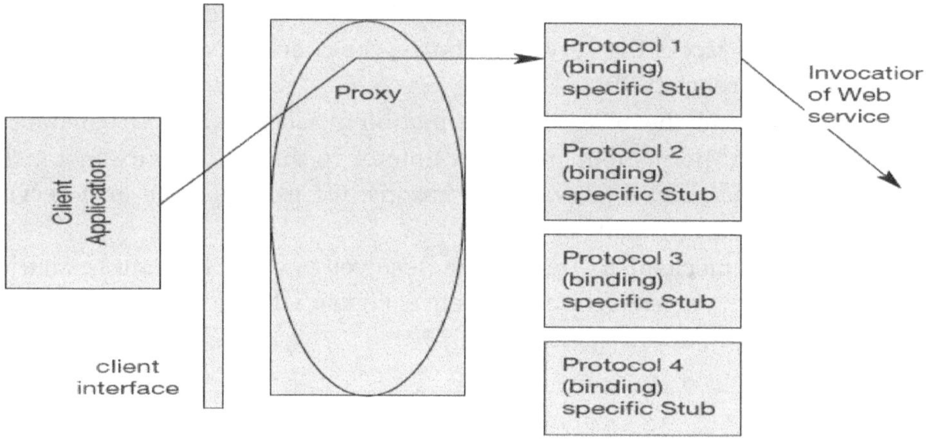

Figure 2.10 Possible client-side runtime architecture

Figure 2.10 depicts possible (but not required) client-side architecture for OGSI.

- In this approach, a clear separation exists between the client application and the client-side representation of the Web service (proxy), including components for marshaling the invocation of a Web service over a chosen binding.
- The client application is insulated from the details of the Web service invocation by a **higher-level abstraction: the client-side interface.**
- This interface is a front end to specific parameter marshaling and message routing that can incorporate various binding options provided by the WSDL. Further, this approach allows certain efficiencies, for example, detecting that the client and the Web service exist on the same network host, therefore avoiding the overhead of preparing for and executing the invocation using network protocols.
- Client application runtime, a proxy provides a client-side representation of remote service instance's interface. **Proxy behaviors** specific to a particular encoding and network protocol (binding, in Web services terminology) are encapsulated in a protocol-specific (binding-specific) stub

Disadvantages

- Not recommended, for developers to build customized code that directly couples client applications to fixed bindings of a particular grid service instance
- This approach introduces significant inflexibility into a system

Stub and client-side infrastructure model

This includes both application-specific services and common infrastructure services that are defined by OGSA. Thus, for most software developers using grid services, the infrastructure and application-level services appear in the form of a class library or programming language interface that is natural to the caller. WSDL and the GWSDL extensions provide support for enabling heterogeneous tools and enabling infrastructure software.

2.2.7.1.3 Client Use of Grid Service Handles and References

- A grid service handle (GSH) can be thought of as a permanent network pointer to a particular grid service instance. The GSH does not provide sufficient information to allow a client to access the service instance; the client needs to "resolve" a GSH into a grid service reference (GSR).
- The GSR contains all the necessary information to access the service instance.
- The GSR is not a "permanent" network pointer to the grid service instance because a GSR may become invalid for various reasons; for example, the grid service instance may be moved to a different server.
- OGSI provides a mechanism, the Handle Resolver to support client resolution of a grid service handle into a grid service reference. Figure 2.11 shows a client application that needs to resolve a GSH into a GSR.

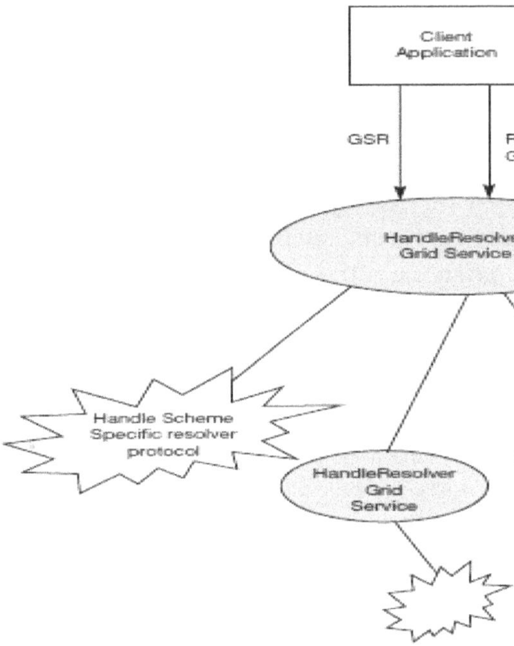

Figure 2.11 Resolving a GSH

HandleResolver

The client resolves a GSH into a GSR by invoking a HandleResolver grid service instance identified by some out-of-band mechanism.

- The HandleResolver can use various means to do the resolution.
- The HandleResolver may have the GSR stored in a local cache.
- The HandleResolver may need to invoke another HandleResolver to resolve the GSH.
- The HandleResolver may use a handle resolution protocol, specified by the particular kind (or scheme) of the GSH to resolve to a GSR.
- The HandleResolver protocol is specific to the kind of GSH being resolved.

2.2.7.1.4 Relationship to Hosting Environment

OGSI does not dictate a particular service-provider-side implementation architecture. A variety of approaches are possible, ranging from implementing the **grid service instance directly as an operating system process** to a sophisticated server-side component model such as J2EE. In the former case, most or even all support for standard grid service behaviors (invocation, lifetime management, registration, etc.) is encapsulated within the user process; for example, via linking with a standard library. In the latter case, many of these behaviors are supported by the hosting environment.

Demarshaling functions

Figure 2.12 illustrates these differences by showing two different approaches to the implementation of argument **demarshaling functions**.

- The invocation message is received at a network protocol termination point (e.g., an HTTP servlet engine) that converts the data in the invocation message into a format consumable by the hosting environment.
- The top part of Figure 2.12 illustrates two grid service instances (the oval) associated with container-managed components (e.g., EJBs within a J2EE container).
- Here, the message is dispatched to these components, with the container frequently providing facilities for demarshaling and decoding the incoming message from a format (such as an XML/SOAP message) into an invocation of the component in native programming language. In some circumstances (the oval), the entire behavior of a grid service instance is completely encapsulated within the component.

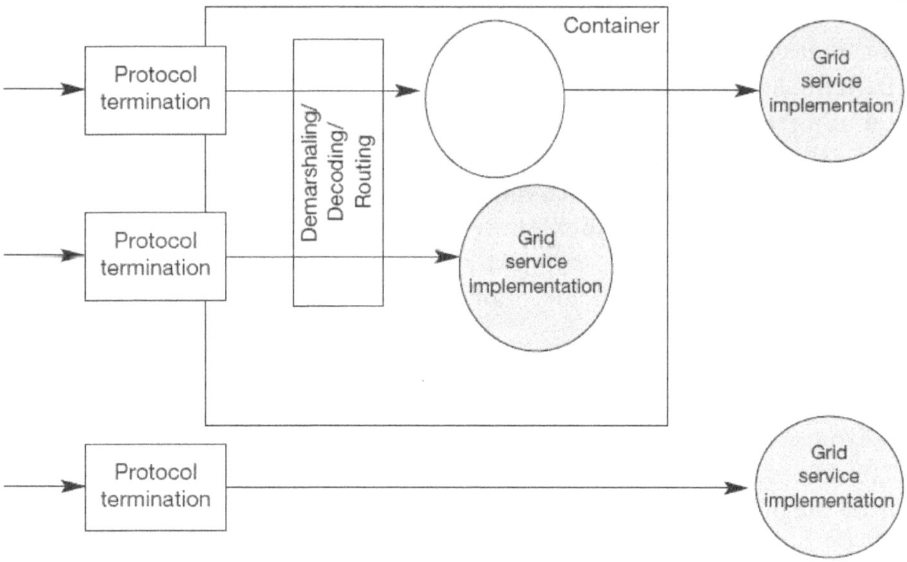

Figure 2.12 Two approaches to the implementation of argument demarshaling functions in a grid service hosting environment

Demarshaling/ decoding

In other cases (the oval), a component will collaborate with other server-side executables, perhaps through an adapter layer, to complete the implementation of the grid service behavior. The bottom part of Figure 2.12 depicts another scenario wherein the entire behavior of the grid service instance, including the **demarshaling/ decoding** of the network message, has been encapsulated within a single executable. Although this approach may have some efficiency advantages, it provides little opportunity for reuse of functionality between grid service implementations.

A container implementation may provide a range of functionality beyond **simple argument demarshaling**. For example, the container implementation may provide lifetime management functions, automatic support for authorization and authentication, request logging, intercepting lifetime management functions, and terminating service instances when a service lifetime expires or an explicit destruction request is received. Thus, one avoids the need to reimplement these common behaviors in different grid service implementations.

2.2.7.3 The Grid Service

The purpose of the OGSI document is to specify the (standardized) interfaces and behaviors that define a grid service. In brief, a grid service is a WSDL-defined service that conforms to a set of conventions relating to its interface definitions and behaviors.

Thus, every grid service is a Web service, though the converse of this statement is not true. The OGSI document expands upon this brief statement by

- Introducing a set of WSDL conventions that one uses in the grid service specification; these conventions have been incorporated in WSDL 1.2.
- Defining service data that provide a standard way for representing and querying metadata and state data from a service instance
- Introducing a series of core properties of grid service, including:
  - Defining grid service description and grid service instance, as organizing principles for their extension and their use.
  - Defining how OGSI models time.
  - Defining the grid service handle and grid service reference constructs that are used to refer to grid service instances.
  - Defining a common approach for conveying fault information from operations. This approach defines a base XML schema definition and associated semantics for WSDL fault messages to support a common interpretation; the approach simply defines the base format for fault messages, without modifying the WSDL fault message model.
  - Defining the life cycle of a grid service instance

2.2.7.4 WSDL Extensions and Conventions

- OGSI is based on Web services; in particular, it uses WSDL as the mechanism to describe the public interfaces of grid services.
- WSDL 1.1 is deficient in two critical areas: **lack of interface (portType) extension and the inability** to describe additional information elements on a portType (lack of open content).
- These deficiencies have been addressed by the W3C Web Services Description Working Group . Because WSDL 1.2 is a "work in progress," OGSI cannot directly incorporate the entire WSDL 1.2 body of work.
- Instead, OGSI defines an extension to WSDL 1.1, isolated to the wsdl:portType element,

which provides the minimal required extensions to WSDL 1.1.

- These extensions to WSDL 1.1 match equivalent functionality agreed to by the W3C Web Services Description Working Group.
- Once WSDL 1.2 [150] is published as a recommendation by the W3C, the Global Grid Forum is committed to defining a follow- on version of OGSI that exploits WSDL 1.2, and to defining a translation from this OGSI v1.0 extension to WSDL 1.2.

2.2.8 Service Data

The approach to **stateful Web services** introduced in OGSI identified the need for a common mechanism to expose a service instance's state data to service requestors for query, update, and change notification. Since this concept is applicable to any Web service including those used outside the context of grid applications, one can propose a common approach to exposing Web service state data called **"serviceData**."

In order to provide a complete description of the interface of a stateful Web service (i.e., a grid service), it is necessary to describe the elements of its state that are externally observable. Service data can be exposed for read, update, or subscription purposes. Since WSDL defines operations and messages for portTypes, the declared state of a service must be externally accessed only through service operations defined as part of the service interface.

Service Data declaration is the mechanism used to express the elements of the publicly available state exposed by the service's interface. ServiceData elements are accessible through operations of the service interfaces such as those defined in this specification. The private internal state of the service instance is not part of the service interface and is therefore not represented through a serviceData declaration.

2.2.8.1 Motivation and Comparison to JavaBean Properties

- The OGSI specification introduces the serviceData concept to provide a flexible, properties style approach to accessing state data of a Web service.
- The serviceData concept is similar to the notion of a public instance variable or field in object-oriented programming languages such as Java, Smalltalk, and C++. ServiceData is similar to JavaBean properties.
  - The JavaBean model defines conventions for method signatures (getXXX/setXXX) to access properties, and helper classes (BeanInfo) to document properties.
- OGSI defines extensible operations for querying (get), updating (set), and subscribing to notification of changes in serviceData elements.
- The serviceDataName element in a GridService portType definition corresponds to the BeanInfo class in JavaBeans. However, OGSI has chosen an XML (WSDL) document that provides information about the serviceData, instead of using a serializable implementation class as in the BeanInfo model.

2.2.8.2 Extending portType with serviceData

- ServiceData defines a new portType child element named serviceData, used to define serviceData elements, or SDEs, associated with that portType.
- These serviceData element definitions are referred to as serviceData declarations, or

SDDs. Initial values for those serviceData elements (marked as "static" serviceData elements) may be specified using the static ServiceData Values element within portType.

- The values of any serviceData element, whether declared statically in the portType or assigned during the life of the
- Web service instance, are called serviceData element values, or SDE values.

2.2.8.3 ServiceDataValues

- Each service instance is associated with a collection of serviceData elements: those serviceData elements defined within the various portTypes that form the service's interface, and also, potentially, additional service-Data elements added at runtime.
- OGSI calls the set of serviceData elements associated with a service instance its "**serviceData set**." A serviceData set may also refer to the set of serviceData elements aggregated from all serviceData elements declared in a portType interface hierarchy.
- Each service instance must have a "logical" XML document, with a root element of serviceDataValues that contains the serviceData element values.
- A service implementation is free to choose how the SDE values are stored.

2.2.8.4 SDE Aggregation within a portType Interface Hierarchy

- WSDL 1.2 has introduced the notion of multiple portType extension, and one can model that construct within the GWSDL namespace.
- A portType can extend zero or more other portTypes.
- There is no direct relationship between a wsdl:service and the portTypes supported by the service modeled in the WSDL syntax.
- The set of portTypes implemented by the service is derived through the port element children of the service element and binding elements referred to from those port elements.
- The serviceData set defined by the service's interface is the set union of the serviceData elements declared in each portType in the complete interface implemented by the service instance.

2.2.8.5 Dynamic serviceData Elements

The grid service portType illustrates the use of dynamic SDEs. This contains a serviceData element named "serviceDataName" that lists the serviceData elements currently defined. This property of a service instance may return a superset of the serviceData elements declared in the GWSDL defining the service interface, allowing the requestor to use the subscribe operation if this serviceDataSet changes, and the findServiceData operation to determine the current serviceDataSet value.

2.2.9 Core Grid Service Properties

This subsection discusses a number of properties and concepts common to all grid services.

2.2.9.1 Service Description and Service Instance

One can distinguish in OGSI between the description of a grid service and an instance of a grid service:

- **A grid service description** describes how a client interacts with service instances. This description is independent of any particular instance. Within a WSDL document, the grid service description is embodied in the most derived portType (i.e., the portType referenced by the wsdl:service element's port children, via referenced binding elements, describing the service) of the instance, along with its associated portTypes (including serviceData declarations), bindings, messages, and types definitions.
- A grid service description may be simultaneously used by any number of **grid service instances**, each of which
  - Embodies some state with which the service description describes how to interact
  - Has one or more grid service handles
  - Has one or more grid service references to it

2.2.9.2 Modeling Time in OGSI

The need arises at various points throughout this specification to represent time that is meaningful to multiple parties in the distributed Grid. For example, information may be tagged by a producer with timestamps in order to convey that information's useful lifetime to consumers. Clients need to negotiate service instance lifetimes with services, and multiple services may need a common understanding of time in order for clients to be able to manage their simultaneous use and interaction.

The **GMT global time standard** is assumed for grid services, allowing operations to refer unambiguously to absolute times. However, assuming the GMT time standard to represent time does not imply any particular level of clock synchronization between clients and services in the grid. In fact, no specific accuracy of synchronization
is specified or expected by OGSI, as this is a service-quality issue.

- Network Time Protocol (NTP) or equivalent function to synchronize their clocks to the global standard GMT time.
- clients and services requiring global ordering or synchronization at a finer granularity than their clock accuracies or resolutions allow for must coordinate through the use of additional synchronization service interfaces, such as through transactions or synthesized global clocks.

In some cases, it is required to represent both zero time and infinite time. Zero time should be represented by a time in the past. However, infinite time requires an extended notion of time. One therefore introduces the following type in the OGSI namespace that may be used in place of xsd: dateTime when a special value of "infinity" is appropriate.

2.2.9.3 XML Element Lifetime Declaration Properties

One can define three XML attributes that together describe the lifetimes associated with an XML element and its subelements. These attributes may be used in any XML element that allows for extensibility attributes, including the serviceData element.

The three lifetimes declaration properties are:

1. **ogsi:goodFrom**. Declares the time from which the content of the element is said to be valid. This is typically the time at which the value was created.
2. **ogsi:goodUntil**. Declares the time until which the content of the element is said to be valid. This property must be greater than or equal to the goodFrom time.
3. **ogsi:availableUntil**. Declares the time until which this element itself is expected to be available, perhaps with updated values. Prior to this time, a client should be able to obtain an updated copy of this element. After this time, a client may no longer be able to get a copy of this element (while still observing cardinality and mutability constraints on this element). This property must be greater than or equal to the goodFrom time.

2.3 Data-Intensive Grid Service Models

Applications in the grid are normally grouped into two categories:

- computation-intensive
- dataintensive
  - For data-intensive applications have to deal with massive amounts of data.
  - The grid system must be specially designed to discover, transfer, and manipulate these massive data sets. Transferring massive data sets is a time-consuming task.
  - Efficient data management demands low-cost storage and high-speed data movement.

Listed in the following paragraphs are several common methods for solving data movement problems.

2.3.1 Data Replication and Unified Namespace

This data access method is also known as **caching**,

- This is often applied to enhance data efficiency in a grid environment.
- By replicating the same data blocks and scattering them in multiple regions of a grid, users can access the same data with locality of references.
- Furthermore, the replicas of the same data set can be a backup for one another.
- Some key data will not be lost in case of failures. However, data replication may demand periodic consistency checks.
- The increase in storage requirements and network bandwidth may cause additional problems.

Replication strategies determine when and where to create a replica of the data.

- The factors to consider include data demand, network conditions, and transfer cost.
- The strategies of replication can be classified into method types:

  ➔ **Dynamic strategies**
  - Can adjust locations and number of data replicas according to changes in conditions (e.g., user behavior).
  - Frequent data-moving operations can result in much more overhead than in static strategies.
  - The replication strategy must be optimized with respect to the status of data replicas.

  - o optimization may be determined based on whether the data replica is being created, deleted, or moved

→ **Static method**

  - o The locations and number of replicas are determined in advance and will not be modified.
  - o Replication operations require little overhead, static strategies cannot adapt to changes in demand, bandwidth, and storage viability.
  - o Optimization is required to determine the location and number of data replicas.

The most common **replication strategies** include preserving locality, minimizing update costs, and maximizing profits.

2.3.2 Grid Data Access Models

- Multiple participants may want to share the same data collection.
  - → To retrieve any piece of data, we need a grid with a unique global namespace.
  - → Similarly, we desire to have unique file names.
- To achieve these, we have to resolve **inconsistencies among multiple data objects** bearing the same name. Access restrictions may be imposed to avoid confusion.
- Data needs to be protected to **avoid leakage and damage**.
- Users who want to access data have to be **authenticated first and then authorized for access**.

In general, there are four access models for organizing a data grid, as listed here and shown in Figure 2.3.

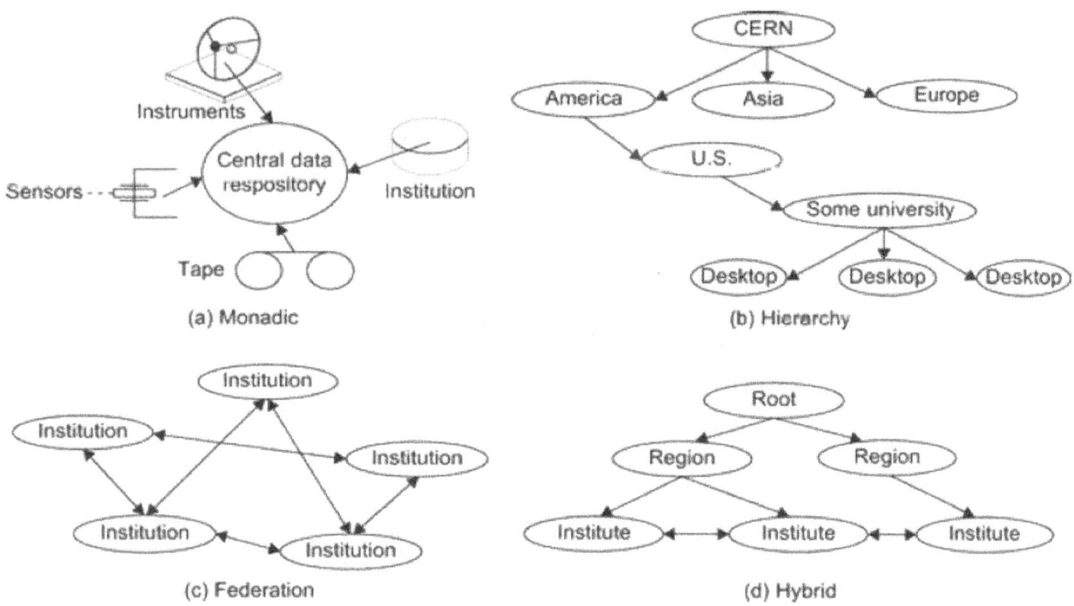

Figure 2.3 Four architectural models for building a data grid

Monadic model: This is a centralized data repository model, shown in Figure 2.3(a).

- All the data is saved in a central data repository.
- When users want to access some data they have to submit requests directly to the central repository.
- No data is replicated for preserving data locality.
- This model is the simplest to implement for a small grid.
- For a large grid, this model is not efficient in terms of performance and reliability.
- Data replication is permitted in this model only when fault tolerance is demanded.

Hierarchical model: The hierarchical model, shown in Figure 2.3(b), is suitable for building a large data grid which has only one large data access directory.

- The data may be transferred from the source to a second-level center.
- Then some data in the regional center is transferred to the third level center.
- After being forwarded several times, specific data objects are accessed directly by users.
- Generally speaking, a higher-level data center has a wider coverage area.
- It provides higher bandwidth for access than a lower-level data center.
- PKI security services are easier to implement in this hierarchical data access model.
- The European Data Grid (EDG) adopts this data access model.

Federation model or mesh model: This data access model shown in Figure 2.3(c) is better suited for designing a data grid with multiple sources of data supplies.

- The data sources are distributed to many different locations.
- Although the data is shared, the data items are still owned and controlled by their original owners. According to predefined access policies, only authenticated users are authorized to request data from any data source.
- This mesh model may cost the most when the number of grid institutions becomes very large.

Hybrid model: This data access model is shown in Figure 2.3(d).

- The model combines the best features of the hierarchical and mesh models.
- Traditional data transfer technology, such as FTP, applies for networks with lower bandwidth.
- Network links in a data grid often have fairly high bandwidth, and other data transfer models are exploited by high-speed data transfer tools such as GridFTP developed with the Globus library.
- The cost of the hybrid model can be traded off between the two extreme models for hierarchical and mesh-connected grids.

2.3.4 Parallel versus Striped Data Transfers

Parallel Data transfer

- Parallel data transfer opens multiple data streams for passing subdivided segments of a file simultaneously.

- The speed of each stream is the same as in sequential streaming, the total time to move data in all streams can be significantly reduced compared to FTP transfer.

Striped data transfer

- Data object is partitioned into a number of sections, and each section is placed in an individual site in a data grid.
- When a user requests this piece of data, a data stream is created for each site, and all the sections of data objects are transferred simultaneously.
- Striped data transfer can utilize the bandwidths of multiple sites more efficiently to speed up data transfer.

2.5 Grid Services and OGSA services

The OGSA is a service-oriented architecture that aims to define a common, standard, and open architecture for grid-based applications. "Open" refers to both the process to develop standards and the standards themselves. In OGSA, everything from registries, to computational tasks, to data resources is considered a service. These extensible set of services are the building blocks of an OGSA-based grid.

OGSA is intended to:

- Facilitate use and management of resources across distributed, heterogeneous environments
- Deliver seamless QoS
- Define open, published interfaces in order to provide interoperability of diverse resources
- Exploit industry-standard integration technologies
- Develop standards that achieve interoperability
- Integrate, virtualize, and manage services and resources in a distributed, heterogeneous Environment
- Deliver functionality as loosely coupled, interacting services aligned with industry-accepted web service standards

Based on OGSA, a grid is built from a small number of standards-based components, called **grid services**.

- Defines a grid service as "a web service that provides a set of well-defined interfaces, following specific conventions (expressed using WSDL)."

OGSA gives a high-level architectural view of **grid services** and doesn't go into much detail when describing grid services. It basically outlines what a grid service should have.

- A grid service implements one or more interfaces, where each interface defines a set of operations that are invoked by exchanging a defined sequence of messages, based on the Open Grid Services Infrastructure (OGSI).

OGSI, also developed by the **Global Grid Forum**, gives a formal and technical specification of a grid service.

Grid service interfaces correspond to **portTypes** in WSDL.

- The set of portTypes supported by a grid service, along with some additional information relating to versioning, are specified in the grid service's serviceType, a WSDL extensibility element defined by OGSA.

- The interfaces address discovery, dynamic service creation, lifetime management, notification, and manageability; whereas the conventions address naming and upgradeability.
- Grid service implementations can target native platform facilities for integration with, and of, existing IT infrastructures.

OGSA services fall into seven broad areas, defined in terms of capabilities frequently required in a grid scenario. Figure 2.4 shows the OGSA architecture.
These services are summarized as follows:

- **Infrastructure Services** Refer to a set of common functionalities, such as naming, typically required by higher level services.
- **Execution Management Services** Concerned with issues such as starting and managing tasks, including placement, provisioning, and life-cycle management. Tasks may range from simple jobs to complex workflows or composite services.
- **Data Management Services** Provide functionality to move data to where it is needed, maintain replicated copies, run queries and updates, and transform data into new formats. These services must handle issues such as data consistency, persistency, and integrity. An OGSA data service is a web service that implements one or more of the base data interfaces to enable access to, and management of, data resources in a distributed environment. The three base interfaces, Data Access, Data Factory, and Data Management, define basic operations for representing, accessing, creating, and managing data.
- **Resource Management Services** Provide management capabilities for grid resources: management of the resources themselves, management of the resources as grid components, and management of the OGSA infrastructure. For example, resources can be monitored, reserved, deployed, and configured as needed to meet application QoS requirements. It also requires an information model (semantics) and data model (representation) of the grid resources and services.
- **Security Services** Facilitate the enforcement of security-related policies within a (virtual) organization, and supports safe resource sharing. Authentication, authorization, and integrity assurance are essential functionalities provided by these services.
- **Information Services** Provide efficient production of, and access to, information about the grid and its constituent resources. The term "information" refers to dynamic data or events used for status monitoring; relatively static data used for discovery; and any data that is logged. Troubleshooting is just one of the possible uses for information provided by these services.
- **Self-Management Services** Support service-level attainment for a set of services (or resources), with as much automation as possible, to reduce the costs and complexity of managing the system.

These services are essential in addressing the increasing complexity of owning and operating an IT infrastructure.

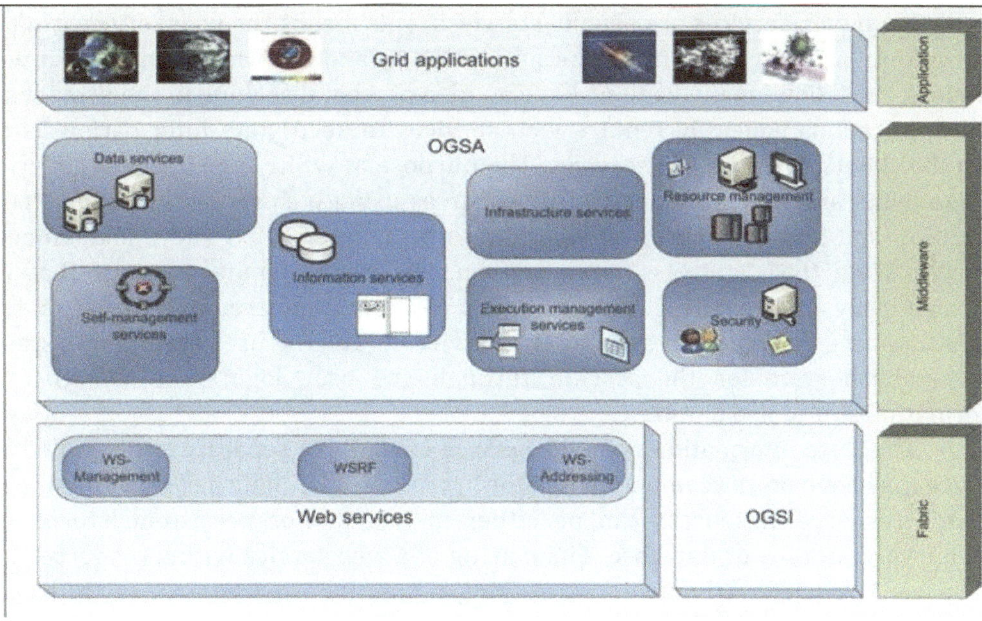

Figure 2.4 The OGSA architecture

OGSA has been adopted as reference grid architecture by a number of grid projects. The first prototype grid service implementation was demonstrated January 29, 2002, at a Globus Toolkit tutorial held at Argonne National Laboratory. Since then, the Globus Toolkit 3.0 and 3.2 have offered an OGSA implementation based on OGSI. Two key properties of a grid service are transience and statefulness. Creation and destruction of a transient grid service can be done dynamically. The creation and lifetime of OGSA grid services are handled following the "factory pattern," . Web service technologies are designed to support loosely coupled, coarse grained dynamic systems, and hence do not meet all grid requirements, such as keeping state information, and thus they are unable to fully address the wide range of distributed systems OGSA is designed to support.

OGSA applies a set of WSDL extensions to represent the identifiers necessary to implement a grid service instance across any system. These extensions were defined by OGSI. A key extension is the grid service reference: a network-wide pointer to a specific grid service instance, which makes that instance accessible to remote client applications. These extensions, including the Grid Service Handle (GSH) and Grid Service Reference (GSR. These extensions include stateful grid services and the shortcomings of OGSI with its dense and long specifications. Further problems concern incompatibility with some current web service tools and the fact that it takes a lot of concepts from object orientation.

Unlike the nature of web services, this has led to close cooperation between the grid and web service communities. As a result of these joint efforts, the Web Services Resource Framework (WSRF) , WS-Addressing , and WS-Notification (WSN) specifications have been proposed to OASIS. Consequently, OGSI extensions to web services have been deprecated in favor of new web service standards, and in particular, WSRF. WSRF is a collection of five different specifications. Of course, they all relate to the management of WS-Resources.

Plain web services are usually stateless. This means the web service can't "remember" information, or keep state, from one invocation to another. However, since a web service is stateless, the following invocations have no idea of what was done in the previous invocations. Grid applications generally require web services to keep state information as they interact with the clients or other web services. The purpose of WSRF is to define a generic framework for modeling and accessing persistent resources using web services in order to facilitate the definition and implementation of a service and the integration and management of multiple services. Note that "stateless" services can, in fact, remember state if that is carried in messages they receive. These could contain a token remembered in a cookie on the client side and a database or cache accessed by the service. Again, the user accessing a stateless service can establish state for the session through the user login that references permanent information stored in a database.

The state information of a web service is kept in a separate entity called a resource. A service may have more than one (singleton) resource, distinguished by assigning a unique key to each resource. Resources can be either in memory or persistent, stored in secondary storage such as a file or database. The pairing of a web service with a resource is called a WS-Resource. The preferred way of addressing a specific WS-Resource is to use the qualified endpoint reference (EPR) construct, proposed by the WS-Addressing specification. Resources store actual data items, referred to as resource properties. Resource properties are usually used to keep service data values, providing information on the current state of the service, or metadata about such values, or they may contain information required to manage the state, such as the time when the resource must be destroyed. Currently, the Globus Toolkit 4.0 provides a set of OGSA capabilities based on WSRF.

3. VIRTUALIZATION

3.1 Cloud Computing and Service Models

Over the past two decades, the world economy has rapidly moved from manufacturing to more service-oriented. Developers of innovative cloud applications no longer acquire large capital equipment in advance. They just rent the resources from some large data centers that have been automated for this purpose. We will study the cloud platform architecture, service models, and programming environments. Users can access and deploy cloud applications from anywhere in the world at very competitive costs. Virtualized cloud platforms are often built on top of large data centers. Clouds aim to power the next generation of data centers by architecting them as virtual resources over automated hardware, databases, user interfaces, and application environments. In this sense, clouds grow out of the desire to build better data centers through automated resource provisioning.

3.1.1 Centralized versus Distributed Computing

Some people argue that cloud computing is centralized computing at data centers. Others claim that cloud computing is the practice of distributed parallel computing over data-center resources. These represent two opposite views of cloud computing. All computations in cloud applications are distributed to servers in a data center. These are mainly virtual machines (VMs) in virtual clusters created out of data-center resources. In this sense, cloud platforms are systems distributed through virtualization.

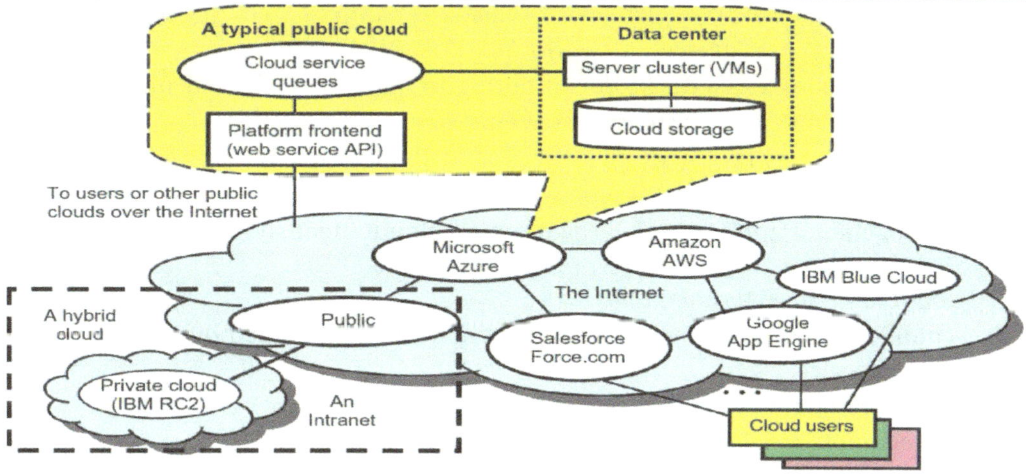

Figure 3.1 Public, private, and hybrid clouds illustrated by functional architecture and connectivity of representative clouds available by 2011.

As Figure 3.1 shows, both public clouds and private clouds are developed in the Internet. As many clouds are generated by commercial providers or by enterprises in a distributed manner, they will be interconnected over the Internet to achieve scalable and efficient computing services. Commercial cloud providers such as Amazon, Google, and Microsoft created their platforms to be distributed geographically.

This distribution is partially attributed to fault tolerance, response latency reduction, and even legal reasons. Intranet-based private clouds are linked to public clouds to get additional resources. Nevertheless, users in Europe may not feel comfortable using clouds in the United States, and vice versa, until extensive service-level agreements (SLAs) are developed between the two user communities.

3.1.2 Cloud deployment models: Public, Private, and Hybrid Clouds

The concept of cloud computing has evolved from cluster, grid, and utility computing. Cluster and grid computing leverage the use of many computers in parallel to solve problems of any size. Utility and Software as a Service (SaaS) provide computing resources as a service with the notion of pay per use.

Cloud Computing

- Leverages dynamic resources to deliver large numbers of services to end users.
- Is a high-throughput computing (HTC) paradigm where the infrastructure provides the services through a large data center or server farms.
- Model enables users to share access to resources from anywhere at any time through their connected devices.
- Avoids large data movement, resulting in much better network bandwidth utilization.
- Machine virtualization has enhanced resource utilization, increased application flexibility, and reduced the total cost of using virtualized data-center resources.
- Offers significant benefit to IT companies by freeing them from the low-level task of setting up the hardware (servers) and managing the system software.
- Applies a virtual platform with elastic resources put together by on-demand provisioning of hardware, software, and data sets, dynamically.
- Main idea is to move desktop computing to a service-oriented platform using server clusters and huge databases at data centers.
- Cloud computing leverages its low cost and simplicity to both providers and users.
- Intends to leverage multitasking to achieve higher throughput by serving many heterogeneous applications, large or small, simultaneously.

3.1.2.1 Public Clouds or External Clouds

➔ A public cloud is built over the Internet and can be accessed by any user who has paid for the service.

➔ Public clouds are owned by service providers and are accessible through a subscription.

➔ A public cloud is hosted, operated, and managed by a third-party vendor from one or more data centers.

- Third-party provider who shares resources and bills on a fine-grained, utility-computing basis.
- Service is offered to multiple customers over a common infrastructure; see Figure 3.2.

➔ The application and infrastructure services are offered on a flexible price-per-use basis

Security Management in public cloud:

- ➔ In a public cloud, security management and day-to-day operations are relegated to the third party vendor, who is responsible for the public cloud service offering.
- ➔ The customer of the public cloud service offering has a low degree of control and failure to notice of the physical and logical security aspects of a private cloud.

Examples

- ➔ Google App Engine (GAE), Amazon Web Services (AWS), Microsoft Azure, IBM Blue Cloud, and Salesforce.com's Force.com

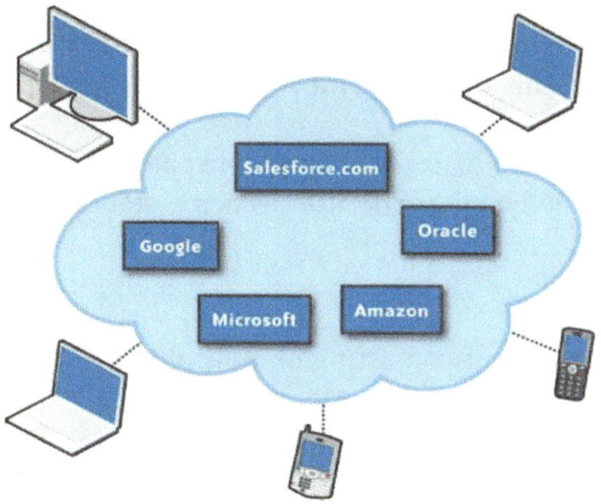

Figure 3.2 public cloud

3.1.2.2 Private Cloud or Internal Cloud

- ➔ A private cloud is built within the domain of an intranet owned by a single organization. Therefore, it is client owned and managed, and its access is limited to the owning clients and their partners.
- ➔ Its deployment was not meant to sell capacity over the Internet through publicly accessible interfaces.
- ➔ Private clouds give local users a flexible and agile private infrastructure to run service workloads
  within their administrative domains.
- ➔ A private cloud is supposed to deliver more efficient and convenient cloud services. It may impact the cloud standardization, while retaining greater customization and organizational control.
- ➔ The organizational customer for a private cloud is responsible for the operation of his private cloud.
- ➔ Private clouds is dedicated to a single organization and is not shared with any other organizations (i.e., the cloud is dedicated to a single organizational tenant).

- ➔ **Types of private cloud patterns**
  - o **Dedicated:** Private clouds hosted within a customer-owned data center or at a collocation facility, and operated by internal IT departments.

- **Community:** Private clouds located at the premises of a third party; owned, managed, and operated by a vendor who is bound by custom SLAs and contractual clauses with security and compliance requirements.
- **Managed:** Private cloud infrastructure owned by a customer and managed by a vendor

Security Management in public cloud

- ➔ In a private cloud operating model, the security management and day-to-day operation of hosts are relegated to internal IT or to a third party with contractual SLAs.
- ➔ In this model, a customer of a private cloud should have a high degree of control and oversight of the physical and logical security aspects of the private cloud infrastructure
- ➔ With that high degree of control and transparency, it is easier for a customer to comply with established corporate security standards, policies, and regulatory compliance.

3.1.2.3 Hybrid Cloud

- ➔ A hybrid cloud environment consisting of multiple internal and/or external providers is a possible deployment for organizations.
- ➔ Private clouds can also support a hybrid cloud model by supplementing local infrastructure with computing capacity from an external public cloud.
    - For example, the Research Compute Cloud (RC2) is a private cloud, built by IBM, that interconnects the computing and IT resources at eight IBM Research Centers scattered throughout the United States, Europe, and Asia.
- ➔ With a hybrid cloud, organizations might run non-core applications in a public cloud, while maintaining core applications and sensitive data in-house in a private cloud (see Figure 3.3).
- ➔ A hybrid cloud provides access to clients, the partner network, and third parties.

Security Management in hybrid cloud

- ➔ Public clouds promote standardization, preserve capital investment, and offer application flexibility. Private clouds attempt to achieve customization and offer higher efficiency, resiliency, security, and privacy. Hybrid clouds operate in the middle, with many compromises in terms of resource sharing.
- ➔ A major concern is to trust that a company's or an individual's information is both secure and private. Establishing this trust is a major milestone in the adoption of the full range of cloud computing.

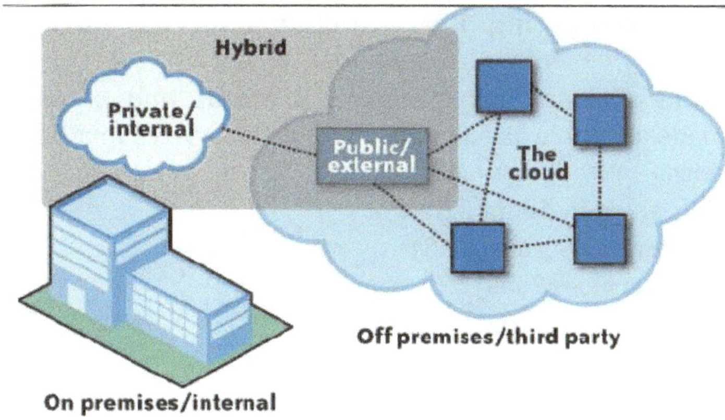

Figure 3.3 Hybrid Cloud

3.1.3 Cloud Design Objectives

Despite the controversy surrounding the replacement of desktop or desk side computing by centralized computing and storage services at data centers or big IT companies, the cloud computing community has reached some consensus on what has to be done to make cloud computing universally acceptable.

The following list highlights six design objectives for cloud computing:

- **Shifting computing from desktops to data centers** Computer processing, storage, and software delivery is shifted away from desktops and local servers and toward data centers over the Internet.
- **Service provisioning and cloud economics** Providers supply cloud services by signing SLAs with consumers and end users. The services must be efficient in terms of computing, storage, and power consumption. Pricing is based on a pay-as-you-go policy.
- **Scalability in performance** The cloud platforms and software and infrastructure services must be able to scale in performance as the number of users increases.
- **Data privacy protection** Can you trust data centers to handle your private data and records?
  This concern must be addressed to make clouds successful as trusted services.
- **High quality of cloud services** The QoS of cloud computing must be standardized to make
  clouds interoperable among multiple providers.
- **New standards and interfaces** This refers to solving the data lock-in problem associated with data centers or cloud providers. Universally accepted APIs and access protocols are needed to provide high portability and flexibility of virtualized applications.

3.2 Categories of cloud computing: Everything as a service: Infrastructure, platform, software

Cloud computing delivers infrastructure, platform, and software (application) as services, which are made available as subscription-based services in a pay-as-you-go model to consumers. The services provided over the cloud can be generally categorized into three different service models given in figure 3.4:

- Infrastructure as a Service (IaaS)
- Platform as a Service (PaaS)
- Software as a Service (SaaS)

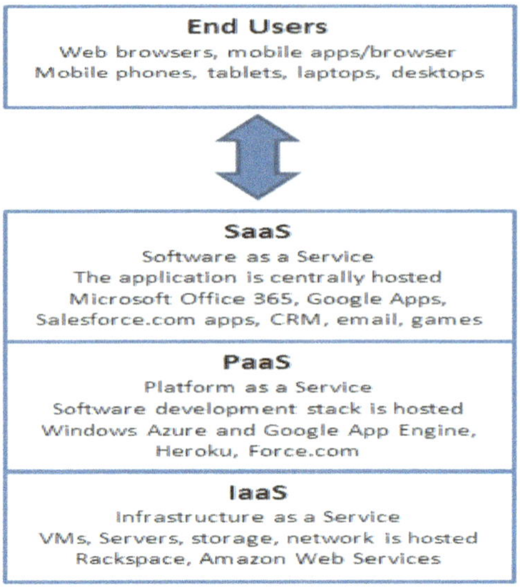

Figure 3.4 Categories of cloud computing

All three models allow users to access services over the Internet, relying entirely on the infrastructures of cloud service providers. These models are offered based on various SLAs between providers and users. In a broad sense, the SLA for cloud computing is addressed in terms of service availability, performance, and data protection and security.

Figure 3.5 illustrates three cloud models at different service levels of the cloud.

- SaaS is applied at the application end using special interfaces by users or clients.
- At the PaaS layer, the cloud platform must perform billing services and handle job queuing, launching, and monitoring services.
- At the bottom layer of the IaaS services, databases, compute instances, the file system, and storage must be provisioned to satisfy user demands.

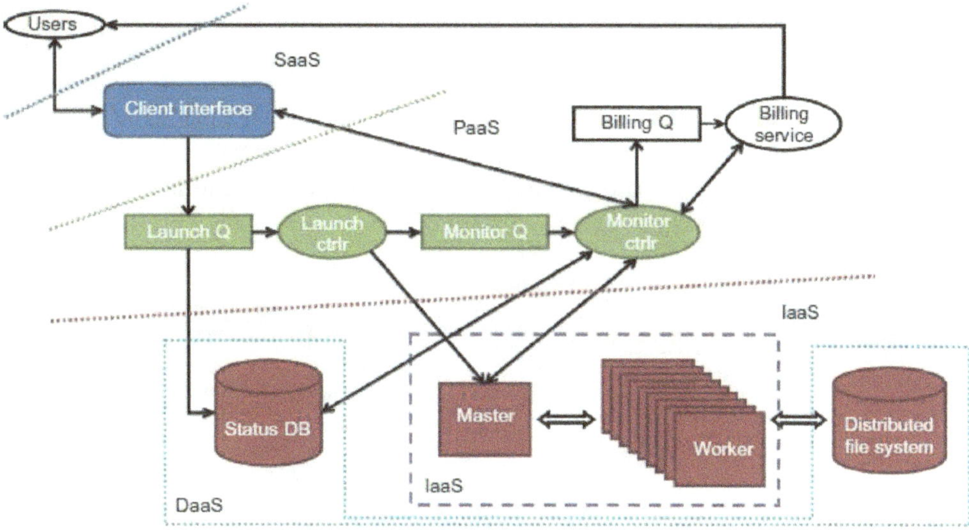

Figure 3.5 The IaaS, PaaS, and SaaS cloud service models at different service levels

3.2.1 Infrastructure-as-a-Service (IaaS)

- This model allows users to use virtualized IT resources for computing, storage, and networking. In short, the service is performed by rented cloud infrastructure.
- The user can deploy and run his applications over his chosen OS environment.
- The user does not manage or control the underlying cloud infrastructure, but has control over the OS, storage, deployed applications, and possibly select networking components.
- This IaaS model encompasses storage as a service, compute instances as a service, and communication as a service.
- The Virtual Private Cloud (VPC) shows how to provide Amazon EC2 clusters and S3 storage to multiple users.

Many startup cloud providers have appeared in recent years. GoGrid, FlexiScale, and Aneka are good examples. Table 4.1 summarizes the IaaS offerings by five public cloud providers

Table 4.1 Public Cloud Offerings of IaaS

Cloud Name	VM Instance capacity	API and Access Tools	Hypervisor, Guest OS
Amazon EC2	Each instance has 1-20 EC2 processor, 1.7-15GB memory, 160-1.69 TB storage	CLI or Web Service Portal	Xen, Linux, Windows
GoGrid	Each instance has 1-6 CPUs, 0.5-8 GB memory, 30-480 GB storage	REST, Java, PHP, Python, Ruby	Xen, Linux, Windows
Rackspace Cloud	Each instance has four core CPU, 0.25-16 GB memory, 10-620 GB storage	REST, Java, PHP, Python, Ruby, C# .NET	Xen, Linux
FlexiScale in the UK	Each instance has 1-4 CPUs, 0.5-16GB memory, 20-270 TB storage	Web Console	Xen, Linux, Windows
Joyent CLoud	Each instance has up to eight CPUs, 0.25-32 GB of memory,30-480 GB storage	No Specific API,SSH, Virtual/Min	OS level Virtualization, Open Solaris

Example IaaS

Amazon VPC for Multiple Tenants

A user can use a private facility for basic computations. When he must meet a specific workload requirement, he can use the Amazon VPC to provide additional EC2 instances or more storage (S3) to handle urgent applications. Figure 3.6 shows VPC which is essentially a private cloud designed to address the privacy concerns of public clouds that hamper their application when sensitive data and software are involved.

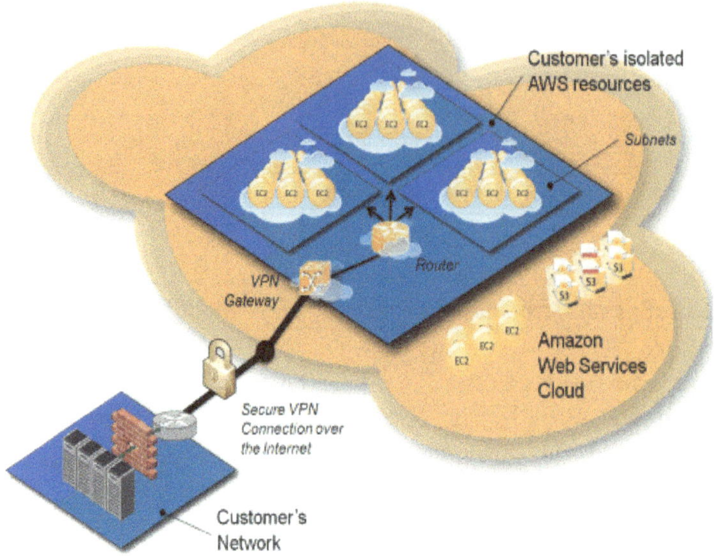

Figure 3.6 Amazon VPC (Virtual Private Cloud)

Amazon EC2 provides the following services: resources from multiple data centers globally distributed, CL1, web services (SOAP and Query), web-based console user interfaces, access to VM instances via SSH and Windows, 99.5 percent available agreements, per-hour pricing, Linux and Windows OSes, and automatic scaling and load balancing. VPC allows the user to isolate provisioned AWS processors, memory, and storage from interference by other users. Both auto-scaling and elastic load balancing services can support related demands. Auto-scaling enables users to automatically scale their VM instance capacity up or down. With auto-scaling, one can ensure that a sufficient number of Amazon EC2 instances are provisioned to meet desired performance. Or one can scale down the VM instance capacity to reduce costs, when the workload is reduced.

3.2.2 Platform as a Service (PaaS)

- PaaS provides a platform allowing customers to develop, run, and manage applications without the complexity of building and maintaining the infrastructure typically associated with developing and launching an app.
- The platform cloud is an integrated computer system consisting of both hardware and software infrastructure.

- The user application can be developed on this virtualized cloud platform using some programming languages and software tools supported by the provider (e.g., Java, Python, .NET).
- The user does not manage the underlying cloud infrastructure.
- The cloud provider supports user application development and testing on a well-defined service platform.
- This PaaS model enables a collaborated software development platform for users from different parts of the world.
- This model also encourages third parties to provide software management, integration, and service monitoring solutions.

Table 3.2 highlights cloud platform services offered by five PaaS services

Table 3.2 Five Public Cloud Offerings of PaaS

Cloud Name	Language an developer Tools	Programming Models supported by Provider	Target Application and Storage Option
Google App Engine	Python, Java, Eclipse-based IDE	MapReduce, Web Programming on demand	Web application and Big Table Storage
Salesforce.com	Apex, Eclipse-based IDE, web based Wizard	Workflow, Excel-like formula, Web programming on demand	Business applications such as CRM
Microsoft Azure	.NET, Azure tools for MS Visual Studio	Unrestricted Model	Enterprise an web applications
Amazon Elastic Map Reduce	Hive, Pig, Cascading, Java, Ruby, Perl, Python, PHP, R, C++	MapReduce	Data processing and e-commerce
Aneka	.NET, stand-alone SDK	Threads, Task, MapReduce	.NET enterprise applications, HPC

Example PaaS

Google App Engine for PaaS Applications

As web applications are running on Google's server clusters, they share the same capability with many other users. The applications have features such as automatic scaling and load balancing which are very convenient while building web applications. The distributed scheduler mechanism can also schedule tasks for triggering events at specified times and regular intervals. Figure 3.7 shows the operational model for GAE. To develop applications using GAE, a development environment must be provided.

Google provides a fully featured local development environment that simulates GAE on the developer's computer. All the functions and application logic can be implemented locally which is quite similar to traditional software development. The coding and debugging stages can be performed locally as well. After these steps are finished, the SDK provided provides a tool for uploading the user's application to Google's infrastructure where the applications are actually deployed. Many additional third-party capabilities, including software management, integration, and service monitoring solutions, are also provided.

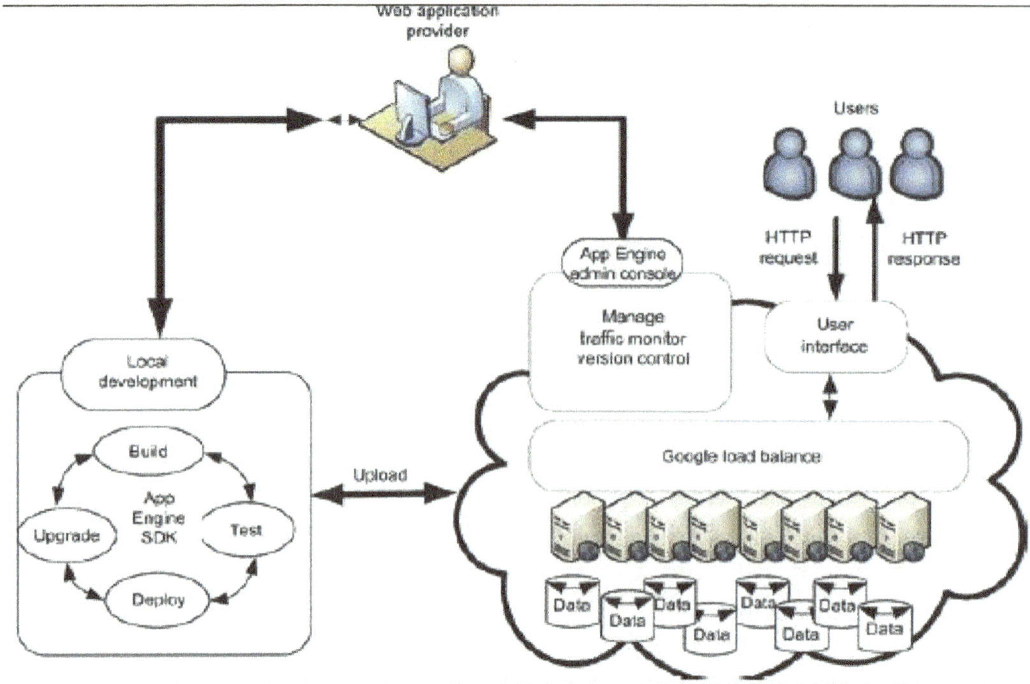

Figure 3.7 Google App Engine platform for PaaS operations

3.2.3 Software as a Service (SaaS)

- The SaaS model provides software applications as a service.
- As a result, on the customer side, there is no upfront investment in servers or software licensing.
- On the provider side, costs are kept rather low, compared with conventional hosting of user applications.
- Customer data is stored in the cloud that is either vendor proprietary or publicly hosted to support PaaS and IaaS.

The best examples of SaaS services include Google Gmail and docs, Microsoft SharePoint, and the CRM software from Salesforce.com. They are all very successful in promoting their own business or are used by thousands of small businesses in their day-to-day operations. Providers such as Google and Microsoft offer integrated IaaS and PaaS services, whereas others such as Amazon and GoGrid offer pure IaaS services and expect third-party PaaS providers such as Manjrasoft to offer application development and deployment services on top of their infrastructure services. To identify important cloud applications in enterprises, the success stories of three real-life cloud applications are presented in Examples for HTC, news media, and business transactions. The benefits of using cloud services are evident in these SaaS applications.

Example:

Three Success Stories on SaaS Applications

1. To discover new drugs through DNA sequence analysis, Eli Lily Company has used Amazon's AWS platform with provisioned server and storage clusters to conduct high-performance biological sequence analysis without using an expensive supercomputer. The benefit of this IaaS application is reduced drug deployment time with much lower costs.
2. The New York Times has applied Amazon's EC2 and S3 services to retrieve useful pictorial information quickly from millions of archival articles and newspapers. The New York Times has significantly reduced the time and cost in getting the job done.
3. Pitney Bowes, an e-commerce company, offers clients the opportunity to perform B2B transactions using the Microsoft Azure platform, along with .NET and SQL services. These offerings have significantly increased the company's client base.

3.2.4 Mashup of Cloud Services

At the time of this writing, public clouds are in use by a growing number of users. Due to the lack of trust in leaking sensitive data in the business world, more and more enterprises, organizations, and communities are developing private clouds that demand deep customization. An enterprise cloud is used by multiple users within an organization. Each user may build some strategic applications on the cloud, and demands customized partitioning of the data, logic, and database in the metadata representation. More private clouds may appear in the future.

Based on a 2010 Google search survey, interest in grid computing is declining rapidly. Cloud mashups have resulted from the need to use multiple clouds simultaneously or in sequence. For example, an industrial supply chain may involve the use of different cloud resources or services at different stages of the chain. Some public repository provides thousands of service APIs and mashups for web commerce services. Popular APIs are provided by Google Maps, Twitter, YouTube, Amazon eCommerce, Salesforce.com, etc.

3.3 Pros and Cons of cloud computing

Advantages of cloud computing

- **No cost on infrastructure**: Cloud computing is divided in 3 major categories as per their services like IaaS, PaaS, SaaS. In all these categories, one thing is common that you don't need to invest on hardware or any infrastructure. In general, every organization has to spend a lot on their IT infrastructure to setup and hire a specialized team. Servers, network devices, ISP connections, storage and software – these are the major things on which you need to invest, if we talk about general IT infrastructure. But if you move to cloud computing services, then you don't need to invest on these. You just go to cloud services provider and buy the cloud service.
- **Minimum management and cost**: Since one doesn't need to invest on the infrastructure, the cost of managing it is also saved. As for IT infrastructure, one needs to hire qualified staff to manage it. While on cloud, the management of its infrastructure is solely of the cloud provider and not of the cloud user, thus again cost saving.

- **Forget about administrative or management hassles**: Whenever there is purchase or upgradation of hardware, a lot of time is wasted looking for best vendors, inviting quotations, negotiating rates, taking approvals, generating POs and waiting for delivery and then in setting up the infrastructure. This whole process includes lots of administrative/managerial tasks that waste a lot of time. While in cloud services, you just need to compare the best cloud service providers and their plans and buy from the one that matches your requirement. And this whole process doesn't take much time and saves a lots of efforts. Your maintenance tasks are also eliminated on cloud.
- **Accessibility and pay per use**: Cloud resources are easily accessible from around the globe – anytime, anywhere and from any device and you have complete access to your resources. This decides your billing also -you only pay for what you use and how much you use. It's like your phone or electricity bill. But on other IT infrastructure, one spends the complete amount in one go and it is very-very rare that those resources are used optimally and thus the investment goes waste.
- **Reliability**: Your infrastructure on cloud increases reliability and availability of applications and services. Cloud services run on pooled and redundant infrastructure which provides you with higher availability of your services.

Disadvantages of Cloud Computing

- **Requires good speed internet with good bandwidth**: To access your cloud services, you need to have a good internet connection always with good bandwidth to upload/download files from/to cloud.
- **Limited control on infrastructure**: Since you are not the owner of infrastructure of cloud, hence you don't have or have a limited access/control on cloud infra.
- **Restricted or limited flexibility**: Although cloud provides a huge list of services but consuming them comes with a lot of restrictions and limited flexibility for your applications or developments.
- **Ongoing costs**: Though you can save your cost of spending on whole infrastructure and its management, but on cloud you need to keep paying for services as long as you use them. But in traditional methods, you only need to invest once.
- **Security**: Security of data is big concern for everyone. Since cloud services are public hence it depends on the provider as to how they are taking care of your data. So, before opting cloud services, it is required that you find a provider who follows max compliances for data security.

3.4 Implementation Levels of Virtualization

Virtualization is a computer architecture technology by which multiple virtual machines (VMs) are **multiplexed** in the same hardware machine. The purpose of a VM is to **enhance resource sharing** by many users and improve computer performance in terms of resource utilization and application flexibility. **Hardware resources** (CPU, memory, I/O devices, etc.) or **software resources** (operating system and software libraries) can be virtualized in various functional layers. The idea is to separate the hardware from the software to yield **better system efficiency**. Virtualization techniques can be applied to **enhance the use of compute engines, networks, and storage**.

In this chapter we will discuss VMs and their applications for building distributed systems. With sufficient storage, any computer platform can be installed in another host computer, even if they use processors with different instruction sets and run with distinct operating systems on the same hardware.

3.4.1 Levels of Virtualization Implementation

A traditional computer runs with a host operating system specially tailored for its hardware architecture, as shown in Figure 3.8(a). After virtualization, different user applications managed by their own operating systems (guest OS) can run on the same hardware, independent of the host OS. This is often done by adding additional software, called a virtualization layer as shown in Figure 3.8(b). This virtualization layer is known as hypervisor or virtual machine monitor (VMM). The VMs are shown in the upper boxes, where applications run with their own guest OS over the virtualized CPU, memory, and I/O resources.

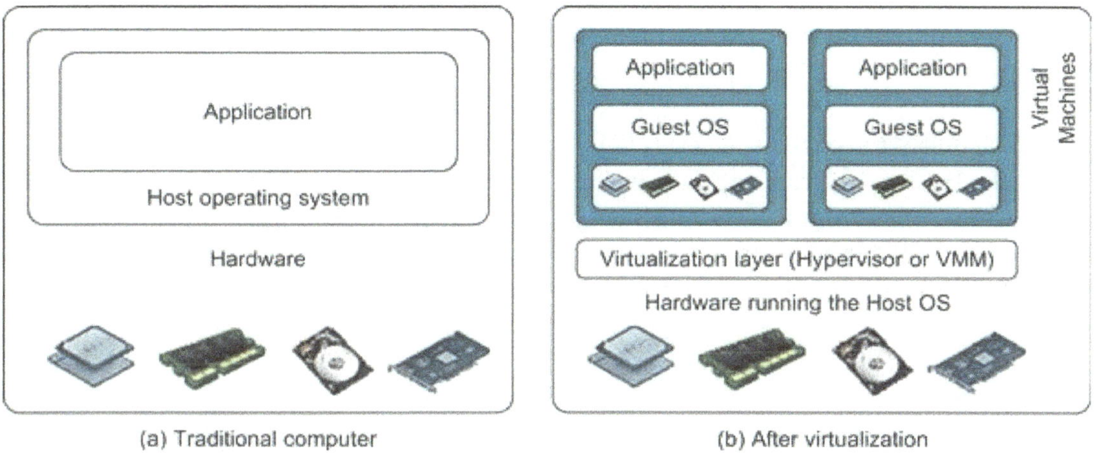

Figure 3.8 The architecture of a computer system before and after virtualization, where VMM stands for virtual machine monitor

The main function of the software layer for virtualization is to virtualize the physical hardware of a host machine into virtual resources to be used by the VMs, exclusively. This can be implemented at various operational levels. The virtualization software creates the abstraction of VMs by interposing a virtualization layer at various levels of a computer system. Common virtualization layers include the instruction set architecture (ISA) level, hardware level, operating system level, library support level, and application level (see Figure 3.9).

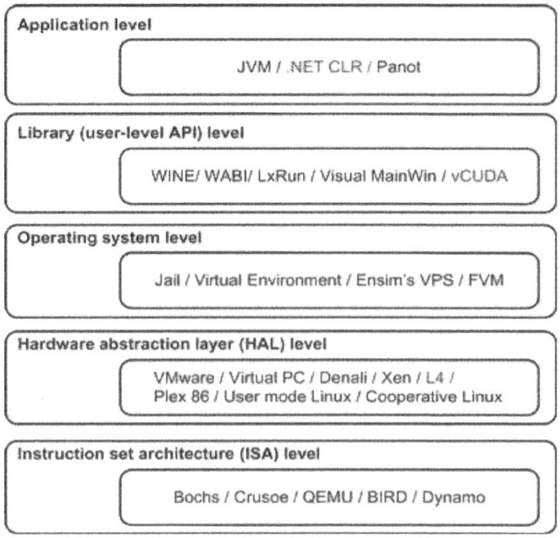

Figure 3.9 Virtualization ranging from hardware to applications in five abstraction levels.

3.4.1.1 Instruction Set Architecture Level (ISA)

- At the ISA level, virtualization is performed by follow a given ISA by the ISA of the host machine. For example, MIPS binary code can run on an x86-based host machine with the help of ISA emulation.
- With this approach, it is possible to run a large amount of legacy binary code written for various processors on any given new hardware host machine.
- Instruction set emulation leads to virtual ISAs created on any hardware machine.

Code interpretation: The basic emulation method is through code interpretation. An interpreter program interprets the source instructions to target instructions one by one. One source instruction may require tens or hundreds of native target instructions to perform its function. Obviously, this process is relatively slow.

Dynamic binary translation: For better performance, dynamic binary translation is used. This approach translates basic blocks of dynamic source instructions to target instructions. The basic blocks can also be extended to program traces or super blocks to increase translation efficiency.

- Instruction set emulation requires binary translation and optimization. A virtual instruction set architecture (V-ISA) thus requires adding a processor-specific software translation layer to the compiler.

3.4.1.2 Hardware Abstraction Level

- Hardware-level virtualization is performed right on top of the bare hardware.
- This approach generates a virtual hardware environment for a VM and the process manages the underlying hardware through virtualization.
- The idea is to virtualize a computer's resources, such as its processors, memory, and I/O devices.

- The intention is to upgrade the hardware utilization rate by multiple users concurrently. For example the Xen hypervisor has been applied to virtualize x86-based machines to run Linux or other guest OS applications.

3.4.1.3 Operating System Level

- This refers to an abstraction layer between traditional OS and user applications.
- OS-level virtualization creates isolated containers on a single physical server and the OS instances to utilize the hardware and software in data centers.
- The containers behave like real servers.
- OS-level virtualization is commonly used in creating virtual hosting environments to allocate hardware resources among a large number of mutually distrusting users.
- It is also used in consolidating server hardware by moving services on separate hosts into containers or VMs on one server.

3.4.1.4 Library Support Level

- Most applications use APIs exported by user-level libraries rather than using lengthy system calls by the OS.
- Virtualization with library interfaces is possible by controlling the communication link between applications and the rest of a system through API hooks.
- The software tool WINE has implemented this approach to support Windows applications on top of UNIX hosts.
- Another example is the vCUDA which allows applications executing within VMs to leverage GPU hardware acceleration.

3.4.1.5 User-Application Level or Process Level Virtualization

Virtualization at the application level virtualizes an application as a VM. On a traditional OS, an application often runs as a process. Therefore, **application-level virtualization** is also known as process-level virtualization. The most popular approach is to deploy high level language (HLL) VMs.

- In this scenario, the virtualization layer sits as an application program on top of the operating system, and the layer exports an abstraction of a VM that can run programs written and compiled to a particular abstract machine definition.
- Any program written in the HLL and compiled for this VM will be able to run on it. The Microsoft .NET CLR and Java Virtual Machine (JVM) are two good examples of this class of VM.
- Other forms of application-level virtualization are known as application isolation, application sandboxing, or application streaming.
- The process involves wrapping the application in a layer that is isolated from the host OS and other applications.
- The result is an application that is much easier to distribute and remove from user workstations. An example is the LANDesk application virtualization platform which deploys software applications as self-contained, executable files in an isolated environment without requiring installation, system modifications, or elevated security privileges.

3.4.1.6 Relative Merits of Different Approaches

Table 3.3 compares the relative merits of implementing virtualization at various levels. The column headings correspond to four technical merits. "Higher Performance" and "Application Flexibility" are self-explanatory. "Implementation Complexity" implies the cost to implement that particular virtualization level. "Application Isolation" refers to the effort required to isolate resources committed to different VMs. Each row corresponds to a particular level of virtualization. Overall, hardware and OS support will yield the highest performance. However, the hardware and application levels are also the most expensive to implement. ISA implementation offers the best application flexibility.

Table 3.3 Relative Merits of Virtualization at Various Levels (More "X"'s Means Higher Merit, with a Maximum of 5 X's)

Level of Implementation	Higher Performance	Application Flexibility	Implementation Complexity	Application Isolation
IAS	X	XXXX	XXX	XXX
Hardware-Level Virtualization	XXXX	XXX	XXXXX	XXXX
OS- Level Virtualization	XXXX	XX	XXX	XX
Runtime library support	XXX	XX	XX	XX
User application level	XX	XX	XXXX	XXXX

3.4.2 VMM Design Requirements and Providers

Virtual Machine Monitor (VMM):

- Hardware-level virtualization inserts a layer between real hardware and traditional operating systems, this layer is commonly called the Virtual Machine Monitor (VMM) and it manages the hardware resources of a computing system.
- Each time programs access the hardware the VMM captures the process that is acts as a traditional OS.
- One hardware component, such as the CPU, can be virtualized as several virtual copies. Therefore, several traditional operating systems which are the same or different can sit on the same set of hardware simultaneously.
- There are three requirements for a VMM.
    - First, a VMM should provide an **environment for programs** which is essentially identical to the original machine.
    - Second, programs run in this environment should show, at worst, **only minor decreases** in speed.
    - Third, a VMM should be in **complete control** of the system resources.
- Any program run under a VMM should exhibit a function identical to that which it runs on the original machine directly.
- Two possible exceptions in terms of differences are permitted with this requirement:

- differences caused by the availability of system resources and
- differences caused by timing dependencies. The former arises when more than one VM is running on the same machine.

A VMM should demonstrate efficiency in using the VMs. Compared with a physical machine, no one prefers a VMM if its efficiency is too low. Traditional emulators and complete software interpreters (simulators) emulate each instruction by means of functions or macros. Such a method provides the most flexible solutions for VMMs. However, emulators or simulators are too slow to be used as real machines. To guarantee the efficiency of a VMM, a statistically dominant subset of the virtual processor's instructions needs to be executed directly by the real processor, with no software intervention by the VMM.

Complete control of these resources by a VMM includes the following aspects:

(1) The VMM is responsible for allocating hardware resources for programs;

(2) It is not possible for a program to access any resource not explicitly allocated to it; and

(3) It is possible under certain circumstances for a VMM to regain control of resources already allocated.

Not all processors satisfy these requirements for a VMM. A VMM is tightly related to the architectures of processors. It is difficult to implement a VMM for some types of processors, such as the x86. Specific limitations include the inability to trap on some privileged instructions. If a processor is not designed to support virtualization primarily, it is necessary to modify the hardware to satisfy the three requirements for a VMM. This is known as **hardware-assisted virtualization**.

3.4.3 Virtualization Support at the OS Level

With the help of VM technology, a new computing mode known as cloud computing is emerging. Cloud computing is transforming the computing landscape by shifting the hardware and staffing costs of managing a computational center to third parties, just like banks.

Cloud computing has at least two challenges:

- The first is the ability to use a variable number of physical machines and VM instances depending on the needs of a problem. For example, a task may need only a single CPU during some phases of execution but may need hundreds of CPUs at other times.
- The second challenge concerns the slow operation of instantiating new VMs. Currently, new VMs originate either as fresh boots or as replicates of a template VM, unaware of the current application state.

3.4.3.1 Why OS-Level Virtualization?

OS-level virtualization provides a feasible solution for these hardware-level virtualization issues.

It is slow to initialize a hardware-level VM because each VM creates its own image from scratch. In a cloud computing environment, perhaps thousands of VMs need to be initialized simultaneously. Besides slow operation, storing the VM images also becomes an issue. As a matter of fact, there is considerable repeated content among VM images. Moreover, full virtualization at the hardware level also has the disadvantages of slow performance and low density, and the need for para-virtualization to modify the guest OS. **To reduce the performance overhead of hardware-level virtualization, even hardware modification is needed.**

Operating system virtualization inserts a virtualization layer inside an operating system to partition a machine's physical resources.

It enables multiple isolated VMs within a single operating system kernel. This kind of VM is often called a *virtual execution environment (VE), Virtual Private System (VPS), or simply container*. From the user's point of view, VEs look like real servers. This means a VE has its own set of processes, file system, user accounts, network interfaces with IP addresses, routing tables, firewall rules, and other personal settings. Although VEs can be customized for different people, they share the same operating system kernel. Therefore, **OS-level virtualization is also called single-OS image virtualization**. Figure 3.10 illustrates operating system virtualization from the point of view of a machine stack.

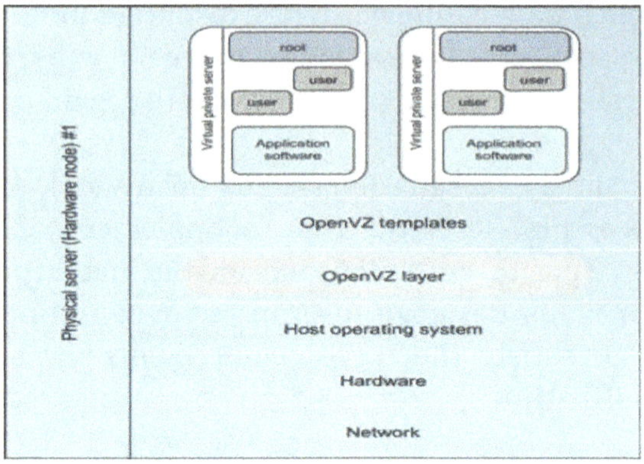

Figure 3.10 The OpenVZ virtualization layer inside the host OS, which provides some OS images to create VMs quickly.

3.4.3.2 Advantages of OS Extensions

Compared to hardware-level virtualization, the benefits of OS extensions are twofold:

(1) VMs at the operating system level have minimal startup/shutdown costs, low resource requirements, and high scalability; and

(2) for an OS-level VM, it is possible for a VM and its host environment to synchronize state changes when necessary.

These benefits can be achieved via two mechanisms of OS-level virtualization:

(1) All OS-level VMs on the same physical machine share a single operating system kernel; and

(2) the virtualization layer can be designed in a way that allows processes in VMs to access as many resources of the host machine as possible, but never to modify them.

In cloud computing, the first and second benefits can be used to overcome the defects of slow initialization of VMs at the hardware level, and being unaware of the current application state, respectively.

3.4.3.3 Disadvantages of OS Extensions

The main disadvantage of OS extensions is that all the VMs at operating system level on a single container must have the same kind of guest operating system.

That is, although different OS-level VMs may have different operating system distributions, they must pertain to the same operating system family. For example, a Windows distribution such as Windows XP cannot run on a Linux based container. However, users of cloud computing have various preferences. Some prefer Windows and others prefer Linux or other operating systems. Therefore, there is a challenge for OS-level virtualization in such cases.

The virtualization layer is inserted inside the OS to partition the hardware resources for multiple VMs to run their applications in multiple virtual environments.

To implement OS-level virtualization, isolated execution environments (VMs) should be created based on a single OS kernel. Furthermore, the access requests from a VM need to be redirected to the VM's local resource partition on the physical machine. For example, the cheroot command in a UNIX system can create several virtual root directories within a host OS. These virtual root directories are the root directories of all VMs created.

There are two ways to implement virtual root directories:

- Duplicating common resources to each VM partition;
- Sharing most resources with the host environment and only creating private resource copies on the VM on demand.

The first way incurs significant resource costs and overhead on a physical machine. This issue neutralizes the benefits of OS-level virtualization, compared with hardware-assisted virtualization. Therefore, OS-level virtualization is often a second choice.

3.4.3.4 Virtualization on Linux or Windows Platforms

Virtualization support on the Windows-based platform is still in the research stage. Most reported OS-level virtualization systems are Linux-based.

- The Linux kernel offers an abstraction layer to allow software processes to work with and operate on resources without knowing the hardware details.
- New hardware may need a new Linux kernel to support.
- Therefore, different Linux platforms use *patched kernels* to provide special support for extended functionality.
- Most Linux platforms are not tied to a special kernel. In such a case, a host can run several VMs simultaneously on the same hardware.
- Table 3.4 summarizes several examples of OS level virtualization tools that have been developed in recent years.

Table 3.4 Virtualization Support for Linux and Windows NT Platforms

Virtualization Support and Source of Information	**Brief Introduction on Functionality and Application Platforms**
Linux vServer for Linux platforms (http://linux-vserver.org/)	Extends Linux kernels to implement a security mechanism to help build VMs by setting resource limits and file attributes and changing the root environment for VM isolation.
OpenVZ for Linux platforms ; http://ftp.openvz.org/doc/OpenVZUsers-Guide.pdf)	Supports virtualization by creating virtual private servers (VPSes); the VPS has its own files, users, process tree, and virtual devices, which can be isolated from other VPSes, and check pointing and live migration are supported

FVM (Feather-Weight Virtual Machines) for virtualizing the Windows NT platforms)	Uses system call interfaces to create VMs at the NY kernel space; multiple VMs are supported by virtualized namespace and copy-on-write

- Two OS tools (Linux vServer and OpenVZ) support Linux platforms to run other platform-based applications through virtualization.

3.1.4 Middleware Support for Virtualization

- Library-level virtualization is also known as *user-level Application Binary Interface (ABI) or API emulation*.
- This type of virtualization can create execution environments for running **alien programs** on a platform rather than creating a VM to run the entire operating system.
- API call interception and remapping are the key functions performed.

Several library level virtualization systems: namely the Windows Application Binary Interface (WABI), lxrun, WINE, Visual MainWin, and vCUDA, which are summarized in Table 3.5.

Table 3.5 Middleware and Library Support for Virtualization

Middleware or Runtime Library and References or Web Link	**Brief Introduction and Application Platforms**
WABI (http://docs.sun.com/app/docs/doc/802-6306)	Middleware that converts Windows system calls running on x86 PCs to Solaris system calls running on SPARC workstations
Lxrun (Linux Run) (http://www.ugcs.caltech.edu/~steven/lxrun/)	A system call emulator that enables Linux applications written for x86 hosts to run on UNIX systems such as the SCO OpenServer
WINE (http://www.winehq.org/)	A library support system for virtualizing x86 processors to run Windows applications under Linux, FreeBSD, and Solaris
VisualMainWin (http://www.mainsoft.com/)	A compiler support system to develop Windows applications using Visual Studio to run on Solaris, Linux, and AIX hosts
vCUDA (IEEE IPDPS 2009)	Virtualization support for using general-purpose GPUs to run data-intensive applications under a special guest OS

- The *WABI* offers middleware to convert Windows system calls to Solaris system calls.
- *Lxrun* is really a system call emulator that enables Linux applications written for x86 hosts to run on UNIX systems. Similarly,
- *Wine* offers library support for virtualizing x86 processors to run Windows applications on UNIX hosts.
- *Visual MainWin* offers a compiler support system to develop Windows applications using Visual Studio to run on some UNIX hosts.
- The *vCUDA* is explained with a graphical illustration in Figure 3.11.

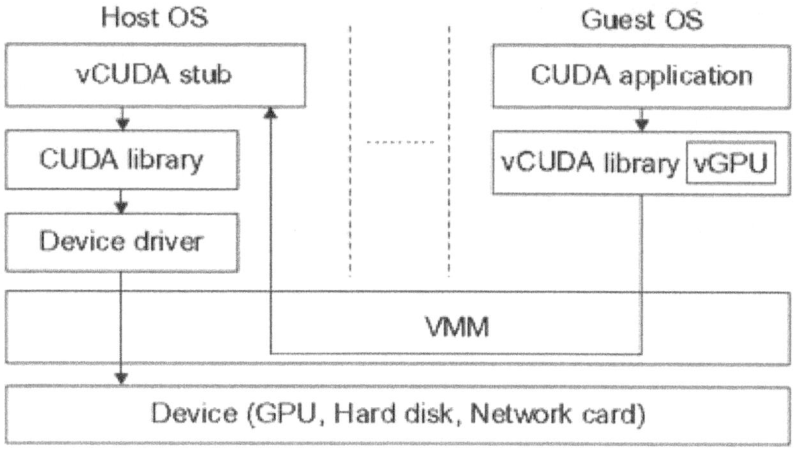

Figure 3.11 Basic concept of the vCUDA architecture

Example: T he vCUDA for Virtualization of General -Purpose GPUs

- CUDA is a programming model and library for general-purpose GPUs. It leverages the high performance of GPUs to run compute-intensive applications on host operating systems.
- vCUDA **virtualizes the CUDA library** and can be installed on guest OSes. When CUDA applications run on a guest OS and issue a call to the CUDA API, vCUDA intercepts the call and redirects it to the CUDA API running on the host OS. Figure 3.11 shows the basic concept of the vCUDA architecture.
- The vCUDA employs a **client-server model** to implement CUDA virtualization. It consists of three user space components: the vCUDA library, a virtual GPU in the guest OS (which acts as a client), and the vCUDA stub in the host OS (which acts as a server).
- The **vCUDA library** resides in the guest OS as a substitute for the standard CUDA library. It is responsible for intercepting and redirecting API calls from the client to the stub.
- Besides these tasks, vCUDA also creates vGPUs and manages them.
  The functionality of a vGPU is threefold:

[1] It abstracts the GPU structure and gives applications a uniform view of the underlying hardware;

[2] when a CUDA application in the guest OS allocates a device's memory the vGPU can return a local virtual address to the application and notify the remote stub to allocate the real device memory,

[3] the vGPU is responsible for storing the CUDA API flow.

- The vCUDA stub receives and interprets remote requests and creates a corresponding execution context for the API calls from the guest OS, then returns the results to the guest OS. The vCUDA stub also manages actual physical resource allocation.

3.5 Virtualization Structures/Tools and Mechanisms

- Before virtualization, the operating system manages the hardware.
- After virtualization, a virtualization layer is inserted between the hardware and the operating system.
- In such a case, the virtualization layer is responsible for converting portions of the real hardware into virtual hardware.
- Therefore, different operating systems such as Linux and Windows can run on the same physical machine, simultaneously.
- Depending on the position of the virtualization layer, there are three typical classes of VM architecture., namely the
  - hypervisor architecture,
  - para-virtualization, and
  - host-based virtualization.

3.5.1 Hypervisor and Xen Architecture

- The hypervisor supports **hardware-level virtualization** on bare metal devices like CPU, memory, disk and network interfaces.
- The hypervisor software sits directly between the physical hardware and its OS. The hypervisor provides **hyper calls** for the guest OSes and applications.
- Hypervisor can assume **micro-kernel architecture** like the Microsoft Hyper-V. or **monolithic hypervisor architecture** like the VMware ESX for server virtualization.

A micro-kernel hypervisor includes only the basic and unchanging functions (such as physical memory management and processor scheduling).

A monolithic hypervisor implements all the aforementioned functions, including those of the device drivers. Therefore, the size of the hypervisor code of a micro-kernel hypervisor is smaller than that of a monolithic hypervisor.

3.5.1.1 The Xen Architecture

Xen

- is an open source hypervisor program developed by Cambridge University.
- is a microkernel hypervisor, which separates the policy from the mechanism.
- hypervisor implements all the mechanisms, leaving the policy to be handled by Domain 0, as shown in Figure 3.12.
- does not include any device drivers natively.
- It just provides a mechanism by which guests OS can have direct access to the physical devices.
- provides a virtual environment located between the hardware and the OS

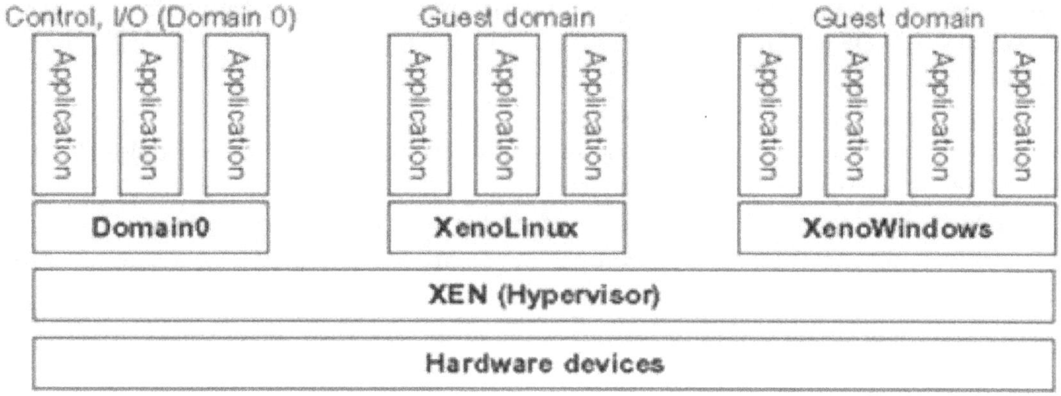

Figure 3.12 The Xen architecture's special domain 0 for control and I/O, and several guest domains for user applications.

The core components of a Xen system are the

- hypervisor,
- kernel, and
- applications.

The organization of the three components is important. Like other virtualization systems, many guest OSes can run on top of the hypervisor. However, not all guest OSes are created equal, and one in particular controls the others. The guest OS, which has control ability, is called Domain 0, and the others are called Domain U. Domain 0 is a privileged guest OS of Xen. It is first loaded when Xen boots without any file system drivers being available. Domain 0 is designed to access hardware directly and manage devices. Therefore, one of the responsibilities of Domain 0 is to allocate and map hardware resources for the guest domains (the Domain U domains).

Traditionally, a machine's lifetime can be envisioned as a straight line where the current state of the machine is a point that progresses monotonically as the software executes. During this time, configuration changes are made, software is installed, and patches are applied. In such an environment, the VM state is akin to a tree: At any point, execution can go into N different branches where multiple instances of a VM can exist at any point in this tree at any given time. VMs are allowed to roll back to previous states in their execution (e.g., to fix configuration errors) or rerun from the same point many times (e.g., as a means of distributing dynamic content or circulating a "live" system image).

3.5.2 Binary Translation with Full Virtualization

Hardware virtualization can be classified into two categories:

- full virtualization and
- Host-based virtualization.

Full virtualization does not need to modify the host OS. It relies on binary translation to trap and to virtualized the execution of certain sensitive, non virtualizable instructions. The guest OSes and their applications consist of noncritical and critical instructions.

Host-based system, both a host OS and a guest OS are used. A virtualization software layer is built between the host OS and guest OS.

3.5.2.1 Full Virtualization

- With full virtualization, noncritical instructions run on the hardware directly while critical instructions are discovered and replaced with traps into the VMM to be emulated by software.
- Both the hypervisor and VMM approaches are considered full virtualization.
- Noncritical instructions do not control hardware or threaten the security of the system, but critical instructions do. Therefore, running noncritical instructions on hardware not only can promote efficiency, but also can ensure system security.

3.5.2.2 Binary Translation of Guest OS Requests Using a VMM

As shown in Figure 3.13, VMware puts the VMM at Ring 0 and the guest OS at Ring 1.

- The VMM scans the instruction stream and identifies the privileged, control and behavior-sensitive instructions. When these instructions are identified, they are trapped into the VMM, which emulates the behavior of these instructions. The method used in this emulation is called **binary translation**.
- Binary translation employs a code cache to store translated hot instructions to improve performance, but it increases the cost of memory usage.
- The performance of full virtualization involves binary translation which is time-consuming, I/O-intensive applications is a really a big challenge.

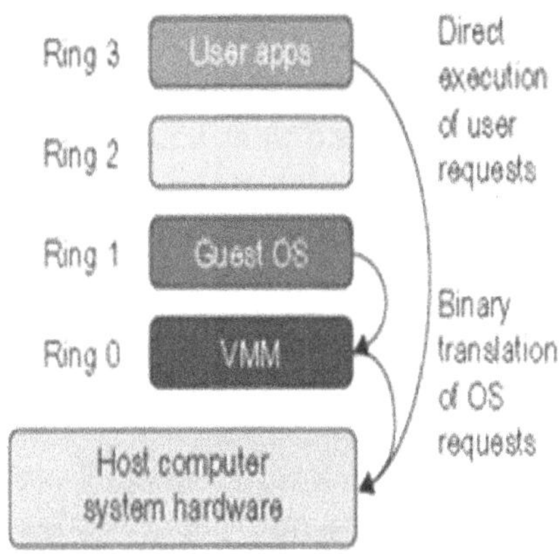

Figure 3.13 Indirect executions of complex instructions via binary translation of guest OS requests using the VMM plus direct execution of simple instructions on the same host.

3.5.2.3 Host-Based Virtualization

- Installs a virtualization layer on top of the host OS which is responsible for managing the hardware.
- Dedicated applications may run on the VMs, other applications can also run with the host OS directly.

This host-based architecture has some distinct advantages,

1. **The user can install this VM architecture without modifying the host OS**. The virtualizing
2. \
3. 2 software can rely on the host OS to provide device drivers and other low-level services. This will simplify the VM design and ease its deployment.
4. **The host-based approach appeals to many host machine configurations**. Compared to the hypervisor/VMM architecture, the performance of the host-based architecture may also be low. When an application requests hardware access, it involves four layers of mapping which downgrades performance significantly. When the ISA of a guest OS is different from the ISA of the underlying hardware, binary translation must be adopted. Although the host-based architecture has flexibility, the performance is too low to be useful in practice.

3.5.3 Para-Virtualization with Compiler Support

- Para-virtualization needs to modify the guest operating systems.
- A para-virtualized VM provides special APIs requiring substantial OS modifications in user applications.
- Para-virtualization attempts to reduce the virtualization overhead, and thus improve performance by modifying only the guest OS kernel.

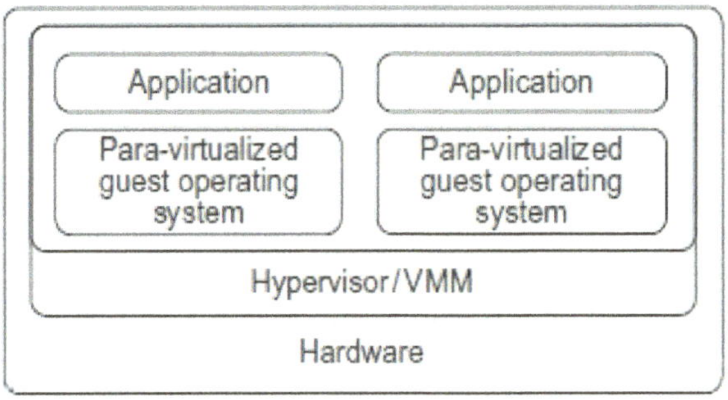

Figure 3.14 Para-virtualized VM architecture, which involves modifying the guest OS kernel to replace nonvirtualizable instructions with hypercalls for the hypervisor or the VMM to carry out the virtualization process.

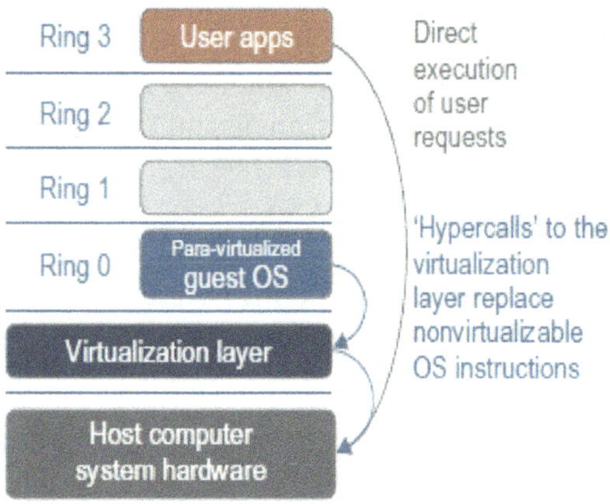

Figure 3.15 The use of a para virtualized guest OS assisted by an intelligent compiler to replace nonvirtualizable OS instructions by hypercalls.

Figure 3.14 illustrates the concept of a para-virtualized VM architecture. The guest operating systems are para-virtualized. They are assisted by an intelligent compiler to replace the nonvirtualizable OS instructions by hypercalls as illustrated in Figure 3.15. The traditional x86 processor offers four instruction execution rings: Rings 0, 1, 2, and 3. The lower the ring number, the higher the privilege of instruction being executed. The OS is responsible for managing the hardware and the privileged instructions to execute at Ring 0, while user-level applications run at Ring 3. The best example of para-virtualization is the KVM to be described below.

3.5.3.1 Para-Virtualization Architecture

When the x86 processor is virtualized, a virtualization layer is inserted between the hardware and the OS. According to the x 86 ring definitions, the virtualization layer should also be installed at Ring 0. Different instructions at Ring 0 may cause some problems. In Figure 3.15, we show that para-virtualization replaces nonvirtualizable instructions with hypercalls that communicate directly with the hypervisor or VMM. However, when the guest OS kernel is modified for virtualization, it can no longer run on the hardware directly.

Pros and Cons of para-virtualization

Para-virtualization reduces the overhead, it has incurred other problems.

- First, its compatibility and portability may be in doubt, because it must support the unmodified OS as well.
- Second, the cost of maintaining para-virtualized OSes is high, because they may require deep OS kernel modifications.
- Finally, the performance advantage of para-virtualization varies greatly due to workload variations.

Compared with full virtualization, para-virtualization is relatively easy and more practical. The main problem in full virtualization is its low performance in binary translation. To speed up binary translation is difficult. Therefore, many virtualization products employ the para-virtualization architecture. The popular Xen, KVM, and VMware ESX are good examples.

3.5.3.2 KVM (Kernel-Based VM)

- This is a Linux para-virtualization system a part of the Linux version 2.6.20 kernel.
- Memory management and scheduling activities are carried out by the existing Linux kernel.
- The KVM does the rest, which makes it simpler than the hypervisor that controls the entire machine.
- KVM is a hardware-assisted para-virtualization tool, which improves performance and supports unmodified guest OSes such as Windows, Linux, Solaris, and other UNIX variants.

3.5.3.3 Para-Virtualization with Compiler Support

- Para-virtualization handles intercepts and emulates privileged and sensitive instructions at compile time.
- The guest OS kernel is modified to replace the privileged and sensitive instructions with hypercalls to the hypervisor or VMM.
- The guest OS running in a guest domain may run at Ring 1 instead of at Ring 0. This implies that the guest OS may not be able to execute some privileged and sensitive instructions.
- The privileged instructions are implemented by hypercalls to the hypervisor.

- After replacing the instructions with hypercalls, the modified guest OS emulates the behavior of the original guest OS.
- On an UNIX system, a system call involves an interrupt or service routine. The hypercalls apply a dedicated service routine in Xen.

Example: VMware ESX Server for Para-Virtualization

VMware pioneered the software market for virtualization. The company has developed virtualization tools for desktop systems and servers as well as virtual infrastructure for large data centers. ESX is a VMM or a hypervisor for bare-metal x86 symmetric multiprocessing (SMP) servers. It accesses hardware resources such as I/O directly and has complete resource management control.

An ESX-enabled server consists of four components: a virtualization layer, a resource manager, hardware interface components, and a service console, as shown in Figure 3.16. To improve performance, the ESX server employs a para-virtualization architecture in which the VM kernel interacts directly with the hardware without involving the host OS.

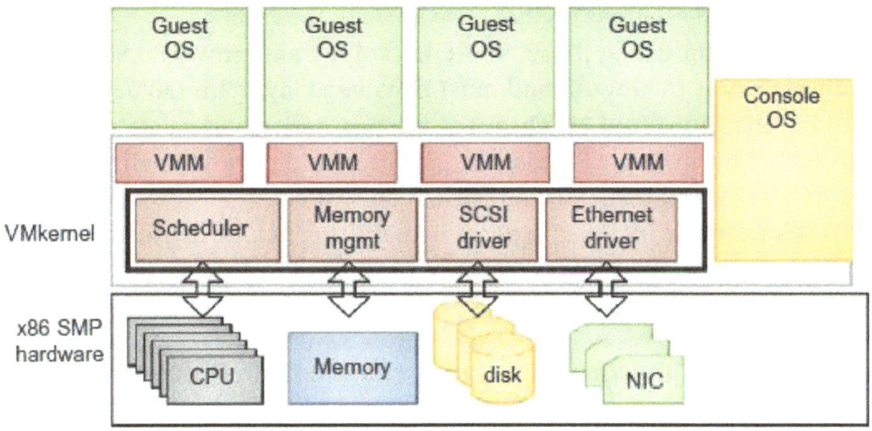

FIGURE 3.16 The VMware ESX server architecture using para-virtualization

The VMM layer virtualizes the physical hardware resources such as CPU, memory, network and disk controllers, and human interface devices. Every VM has its own set of virtual hardware resources. The resource manager allocates CPU, memory disk, and network bandwidth and maps them to the virtual hardware resource set of each VM created. Hardware interface components are the device drivers and the VMware ESX Server File System. The service console is responsible for booting the system, initiating the execution of the VMM and resource manager, and relinquishing control to those layers. It also facilitates the process for system administrators.

3.6 Virtualization of CPU, Memory, and I/O Devices

To support virtualization, x86 employ a special running mode and instructions, known as **hardware-assisted virtualization**. In this way, the VMM and guest OS run in different modes and all sensitive instructions of the guest OS and its applications are trapped in the VMM. To save processor states, mode switching is completed by hardware.

3.6.1 Hardware Support for Virtualization

Modern operating systems and processors permit multiple processes to run simultaneously. If there is no protection mechanism in a processor, all instructions from different processes will access the hardware directly and cause a system crash. Therefore, all processors have at least two modes, **user mode and supervisor mode**, to ensure controlled access of critical hardware. Instructions running in supervisor mode are called **privileged instructions**. Other instructions are **unprivileged instructions**.

In a virtualized environment, it is more difficult to make OSes and applications run correctly because there are more layers in the machine stack.

- The **VMware Workstation** is a VM software suite for x86 and x86-64 computers. This software suite allows users to set up multiple x86 and x86-64 virtual computers and to use one or more of these VMs simultaneously with the host operating system. The VMware Workstation assumes the host-based virtualization.
- **Xen** is a hypervisor for use in IA-32, x86-64, Itanium, and PowerPC 970 hosts. Actually, Xen modifies Linux as the lowest and most privileged layer, or a hypervisor.
- **KVM (Kernel-based Virtual Machine)** is a Linux kernel virtualization infrastructure. KVM can support hardware-assisted virtualization and para-virtualization by using the Intel VT-x or AMD-v and VirtIO framework, respectively.
- The **VirtIO framework** includes a para-virtual Ethernet card, a disk I/O controller, a balloon device for adjusting guest memory usage, and a VGA graphics interface using VMware drivers.

Example: Hardware Support for Virtualization in the Intel x86 Processor

Since software-based virtualization techniques are complicated and incur performance overhead, Intel provides a hardware-assist technique to make virtualization easy and improve performance. Figure 3.17 provides an overview of Intel's full virtualization techniques. For processor virtualization, Intel offers the VT-x or VT-i technique. VT-x adds a privileged mode (VMX Root Mode) and some instructions to processors. This enhancement traps all sensitive instructions in the VMM automatically. For memory virtualization, Intel offers the EPT, which translates the virtual address to the machine's physical addresses to improve performance. For I/O virtualization, Intel implements VT-d and VT-c to support this.

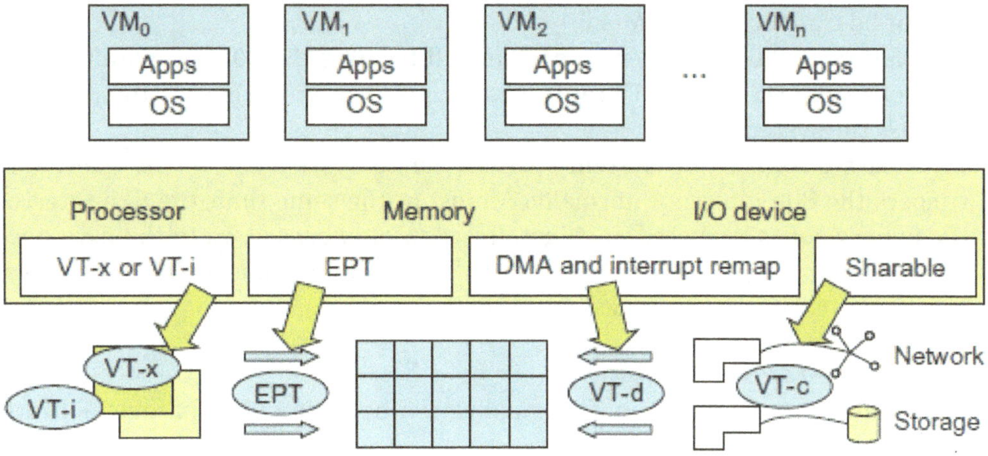

Figure 3.17 Intel hardware support for virtualization of processor, memory, and I/O devices.

3.6.2 CPU Virtualization

A VM is a duplicate of an existing computer system in which a majority of the VM instructions are executed on the host processor in native mode. Thus, unprivileged instructions of VMs run directly on the host machine for **higher efficiency**. Other critical instructions should be handled carefully for **correctness and stability**.

The critical instructions are divided into three categories:

- **privileged instructions**- execute in a privileged mode and will be trapped if executed outside this mode
- **Control sensitive instructions**- attempt to change the configuration of resources used.
- **Behavior-sensitive instructions**- have different behaviors depending on the configuration of resources, including the load and store operations over the virtual memory.

CPU architecture is virtualizable if it supports the ability to run the **VM's privileged and unprivileged instructions** in the CPU's user mode while the VMM runs in supervisor mode.

When the privileged instructions like control and behavior sensitive instructions of a VM are executed, they are trapped in the VMM.

- In this case, the VMM acts as a **unified mediator** for hardware access from different VMs to guarantee the correctness and stability of the whole system.

Example:

- o RISC CPU architectures can be naturally virtualized because all control- and behavior-sensitive instructions are privileged instructions.
- o x86 CPU architectures are not primarily designed to support virtualization. This is because about 10 sensitive instructions, such as SGDT and SMSW, are not

privileged instructions. When these instructions execute in virtualization, they cannot be trapped in the VMM.

- On a native UNIX-like system, a system call triggers the 80h interrupt and passes control to the OS kernel. The interrupt handler in the kernel is then invoked to process the system call.
- On a para-virtualization system such as Xen, a system call in the guest OS first triggers the 80h interrupt normally. Almost at the same time, the 82h interrupt in the hypervisor is triggered. Incidentally, control is passed on to the hypervisor as well. When the hypervisor completes its task for the guest OS system call, it passes control back to the guest OS kernel.

3.6.2.1 Hardware-Assisted CPU Virtualization

- This technique attempts to simplify virtualization because full or paravirtualization is complicated.
- Intel and AMD add an additional mode called privilege mode level to x86 processors.
- All the privileged and sensitive instructions are trapped in the hypervisor automatically.
- This technique removes the difficulty of implementing binary translation of full virtualization. It also lets the operating system run in VMs without modification.

Example Intel Hardware-Assisted CPU Virtualization

Although x86 processors are not virtualizable primarily, great effort is taken to virtualize them. They are used widely in comparing RISC processors that the bulk of x86-based legacy systems cannot discard easily. Virtualization of x86 processors is detailed in the following sections. Intel's VT-x technology is an example of hardware-assisted virtualization, as shown in Figure 3.18. Intel calls the privilege level of x86 processors the VMX Root Mode. In order to control the start and stop of a VM and allocate a memory page to maintain the CPU state for VMs, a set of additional instructions is added. At the time of this writing, Xen, VMware, and the Microsoft Virtual PC all implement their hypervisors by using the VT-x technology.

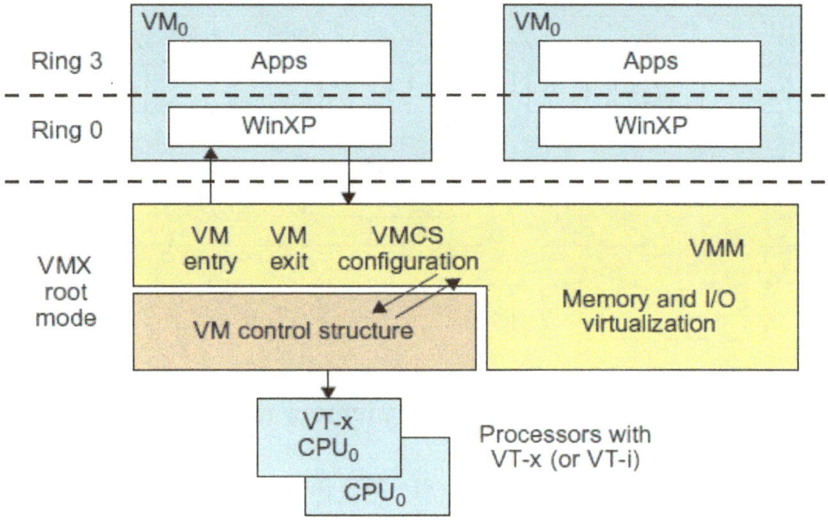

Figure 3.18 Intel hardware-assisted CPU virtualization

Generally, hardware-assisted virtualization should have high efficiency. However, since the transition from the hypervisor to the guest OS incurs high overhead switches between processor modes, it sometimes cannot outperform binary translation. Hence, virtualization systems such as VMware now use a hybrid approach, in which a few tasks are offloaded to the hardware but the rest is still done in software. In addition, para-virtualization and hardware-assisted virtualization can be combined to improve the performance further.

3.6.3 Memory Virtualization

- Virtual memory virtualization is similar to the virtual memory support provided by modern operating systems.
- In a virtual execution environment, virtual memory virtualization involves sharing the physical system memory in RAM and dynamically allocating it to the physical memory of the VMs.
- That means a two-stage mapping process should be maintained by the guest OS and the VMM, respectively: Virtual memory to physical memory and physical memory to machine memory.
- The guest OS continues to control the mapping of virtual addresses to the physical memory addresses of VMs.
- The guest OS cannot directly access the actual machine memory.
- The VMM is responsible for mapping the guest physical memory to the actual machine memory.

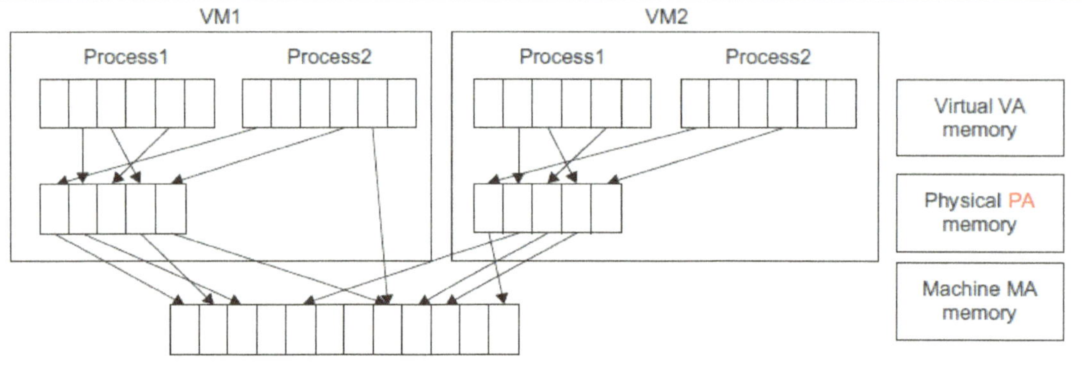

Figure 3.19 Two-level memory mapping procedure.

Figure 3.19 shows the two-level memory mapping procedure

[1] Each page table of the guest OSes has a separate page table in the VMM corresponding to it, the VMM page table is called the **shadow page table**. The MMU handles **virtual-to-physical** translations as defined by the OS.

[2] The **physical memory addresses are translated to machine addresses** using another set of page tables defined by the hypervisor. Modern operating systems maintain a set of page tables for every process, the shadow page tables will get **flooded**, the performance overhead and cost of memory will be very high.

- VMware uses shadow page tables to perform virtual-memory-to-machine-memory address translation. Processors use TLB hardware to map the virtual memory directly to the machine memory to avoid the two levels of translation on every access.

When the guest OS changes the virtual memory to a physical memory mapping, the VMM updates the shadow page tables to enable a direct lookup.

- The AMD Barcelona processor has featured hardware-assisted memory virtualization since 2007.

It provides hardware assistance to the two-stage address translation in a virtual execution environment by using a technology called nested paging.

Example: Extended Page Table by Intel for Memory Virtualization

Since the efficiency of the software shadow page table technique was too low, Intel developed a hardware-based EPT technique to improve it, as illustrated in Figure 3.20. In addition, Intel offers a Virtual Processor ID (VPID) to improve use of the TLB. Therefore, the performance of memory virtualization is greatly improved. In Figure 3.20, the page tables of the guest OS and EPT are all four-level.

When a virtual address needs to be translated, the CPU will first look for the L4 page table pointed to by Guest CR3. Since the address in Guest CR3 is a physical address in the guest OS, the CPU needs to convert the Guest CR3 GPA to the host physical address (HPA) using EPT. In this procedure, the CPU will check the EPT TLB to see if the translation is there. If there is no required translation in the EPT TLB, the CPU will look for it in the EPT. If the CPU cannot find the translation in the EPT, an EPT violation exception will be raised.

When the GPA of the L4 page table is obtained, the CPU will calculate the GPA of the L3 page table by using the GVA and the content of the L4 page table. If the entry corresponding to the GVA in the L4 page table is a page fault, the CPU will generate a page fault interrupt and will let the guest OS kernel handle the interrupt. When the PGA of the L3 page table is obtained, the CPU will look for the EPT to get the HPA of the L3 page table, as described earlier. To get the HPA corresponding to a GVA, the CPU needs to look for the EPT five times, and each time, the memory needs to be accessed four times. Therefore, there are 20 memory accesses in the worst case, which is still very slow. To overcome this shortcoming, Intel increased the size of the EPT TLB to decrease the number of memory accesses.

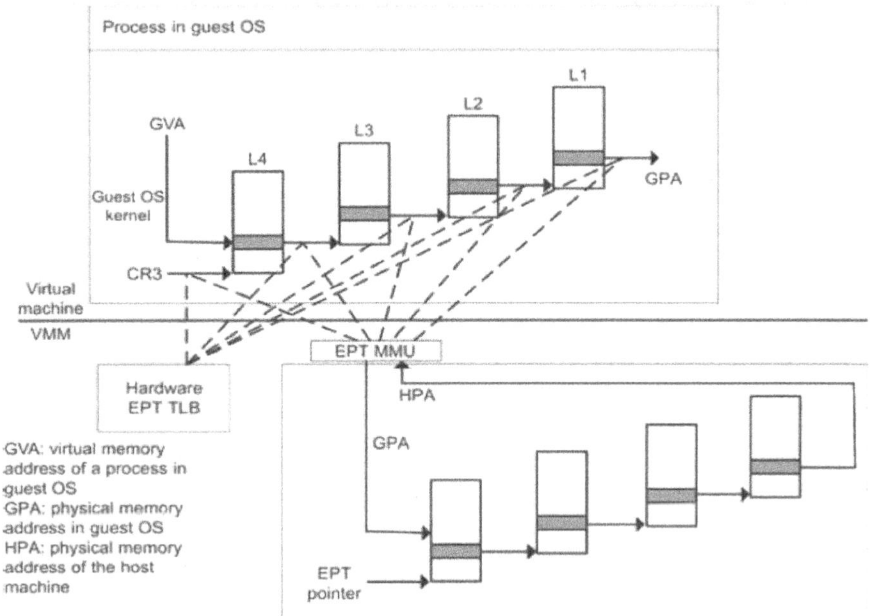

Figure 3.20 Memory virtualization using EPT by Intel , the EPT is also known as the shadow page table

3.6.4 I/O Virtualization

I/O virtualization involves managing the routing of I/O requests between virtual devices and the shared physical hardware.

There are three ways to implement I/O virtualization:

- full device emulation,
- para-virtualization, and
- direct I/O

Full device emulation

- This approach emulates well-known, real-world devices.
- All the functions of a device or bus infrastructure, such as device enumeration, identification, interrupts, and DMA, are replicated in software.
- This software is located in the VMM and acts as a virtual device. The I/O access requests of the guest OS are trapped in the VMM which interacts with the I/O devices.

- The full device emulation approach is shown in Figure 3.21.
- A single hardware device can be shared by multiple VMs that run concurrently.

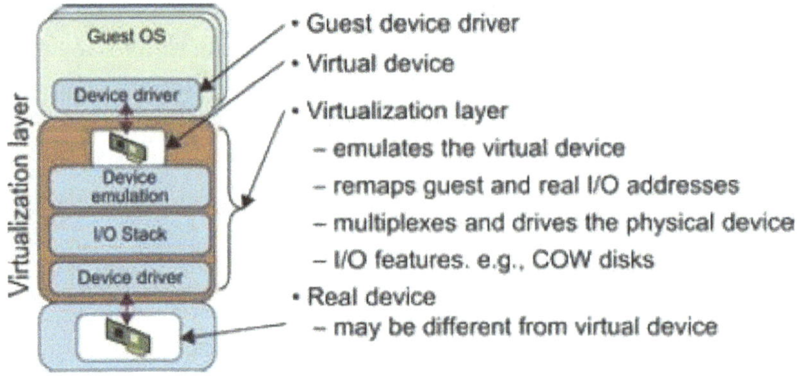

Figure 3.21 Device emulation for I/O virtualization implemented inside the middle layer that maps real I/O devices into the virtual devices for the guest device driver to use.

The para-virtualization method or split driver model

- Consisting of a frontend driver and a backend driver. The frontend driver is running in Domain U and the backend driver is running in Domain 0.
- They interact with each other via a block of shared memory.
- The frontend driver manages the I/O requests of the guest OSes and the backend driver is responsible for managing the real I/O devices and multiplexing the I/O data of different VMs.
- Although para-I/O virtualization achieves better device performance than full device emulation, it comes with a higher CPU overhead.

Direct I/O virtualization

- The VM access devices directly.
- It can achieve close-to-native performance without high CPU costs.
- There are a lot of challenges for commodity hardware devices.

For example, when a physical device is reclaimed (required by workload migration) for later reassignment, it may have been set to an arbitrary state (e.g., DMA to some arbitrary memory locations) that can function incorrectly or even crash the whole system.

- Hardware-assisted I/O virtualization is critical.
  Intel VT-d supports the remapping of I/O DMA transfers and device-generated interrupts. The architecture of VT-d provides the flexibility to support multiple usage models that may run unmodified, special-purpose, or "virtualization-aware" guest OSes.

Self-virtualized I/O (SV-IO)

- The key idea of SVIO is to harness the rich resources of a multicore processor. All tasks associated with virtualizing an I/O device are encapsulated in SV-IO.

- It provides virtual devices and an associated access API to VMs and a management API to the VMM. SV-IO defines one virtual interface (VIF) for every kind of virtualized I/O device, such as virtual network interfaces, virtual block devices (disk), virtual camera devices, and others.
- The guest OS interacts with the VIFs via VIF device drivers.
- Each VIF consists of two message queues. One is for outgoing messages to the devices and the other is for incoming messages from the devices.
- Each VIF has a unique ID for identifying it in SV-IO.

Example: VMware Workstation for I/O Virtualization

The VMware Workstation runs as an application. It leverages the I/O device support in guest OSes, host OSes, and VMM to implement I/O virtualization. The application portion (VMApp) uses a driver loaded into the host operating system (VMDriver) to establish the privileged VMM, which runs directly on the hardware. A given physical processor is executed in either the host world or the VMM world, with the VMDriver facilitating the transfer of control between the two worlds. The VMware Workstation employs full device emulation to implement I/O virtualization. Figure 3.22 shows the functional blocks used in sending and receiving packets via the emulated virtual NIC.

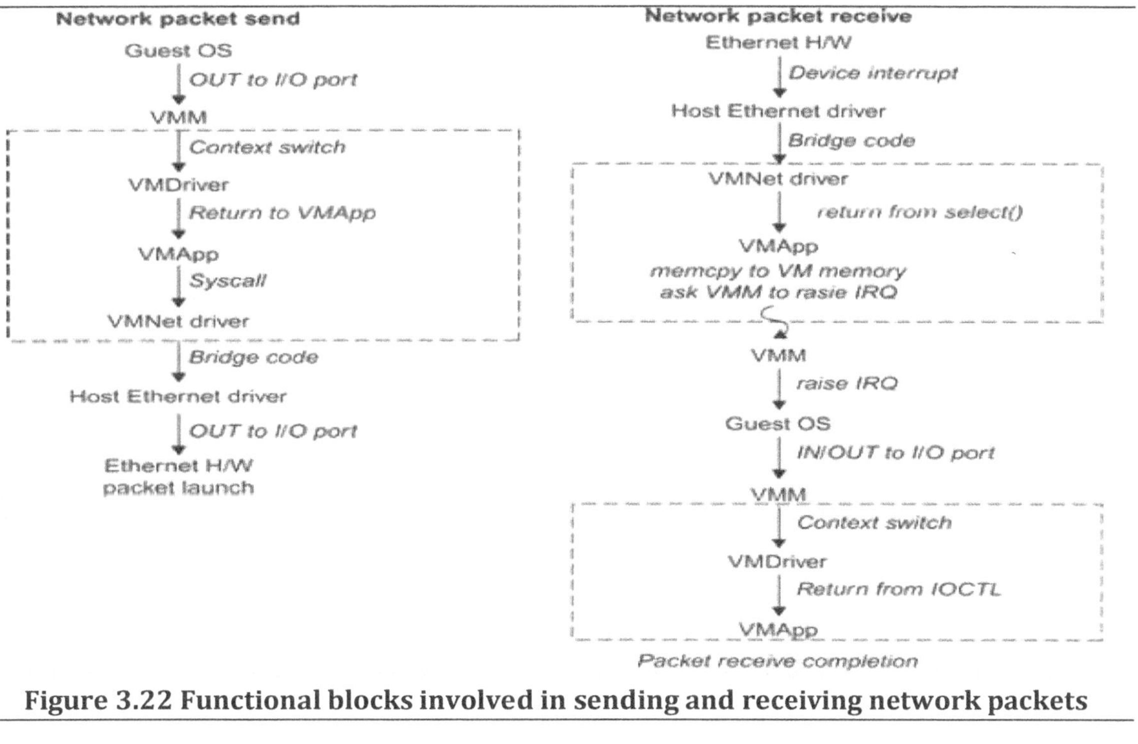

Figure 3.22 Functional blocks involved in sending and receiving network packets

The virtual NIC models an AMD Lance Am79C970A controller. The device driver for a Lance controller in the guest OS initiates packet transmissions by reading and writing a sequence of virtual I/O ports; each read or write switches back to the VMApp to emulate the Lance port accesses. When the last OUT instruction of the sequence is encountered, the Lance emulator calls a normal write() to the VMNet driver. The VMNet driver then passes the packet onto the network via a host NIC and then the VMApp switches back to the VMM. The switch raises a virtual interrupt to notify the guest device driver that the packet was sent. Packet receives occur in reverse.

3.7 Virtual Clusters and Resource Management

A physical cluster is a collection of servers (physical machines) interconnected by a physical network such as a LAN. we will study three critical design issues of virtual clusters: live migration of VMs, memory and file migrations, and dynamic deployment of virtual clusters.

When a traditional VM is initialized, the administrator needs to manually write configuration information or specify the configuration sources. When more VMs join a network, an inefficient configuration always causes problems with overloading or underutilization. Amazon's Elastic Compute Cloud (EC2) is a good example of a web service that provides elastic computing power in a cloud. EC2 permits customers to create VMs and to manage user accounts over the time of their use.

Most virtualization platforms, including XenServer and VMware ESX Server, support a bridging mode which allows all domains to appear on the network as individual hosts. By using this mode, VMs can communicate with one another freely through the virtual network interface card and configure the network automatically.

3.7.1 Physical versus Virtual Clusters

Virtual clusters are built with VMs installed at distributed servers from one or more physical clusters. The VMs in a virtual cluster are interconnected logically by a virtual network across several physical networks. Figure 3.23 illustrates the concepts of virtual clusters and physical clusters. Each virtual cluster is formed with physical machines or a VM hosted by multiple physical clusters. The virtual cluster boundaries are shown as distinct boundaries.

The provisioning of VMs to a virtual cluster is done dynamically to have the following interesting properties:

- The virtual cluster nodes can be either physical or virtual machines. Multiple VMs running with different OSes can be deployed on the same physical node.
- A VM runs with a guest OS, which is often different from the host OS, that manages the resources in the physical machine, where the VM is implemented.
- The purpose of using VMs is to consolidate multiple functionalities on the same server. This will greatly enhance server utilization and application flexibility.
- VMs can be colonized (replicated) in multiple servers for the purpose of promoting distributed parallelism, fault tolerance, and disaster recovery.
- The size (number of nodes) of a virtual cluster can grow or shrink dynamically, similar to the way an overlay network varies in size in a peer-to-peer (P2P) network.
- The failure of any physical nodes may disable some VMs installed on the failing nodes. But the failure of VMs will not pull down the host system.

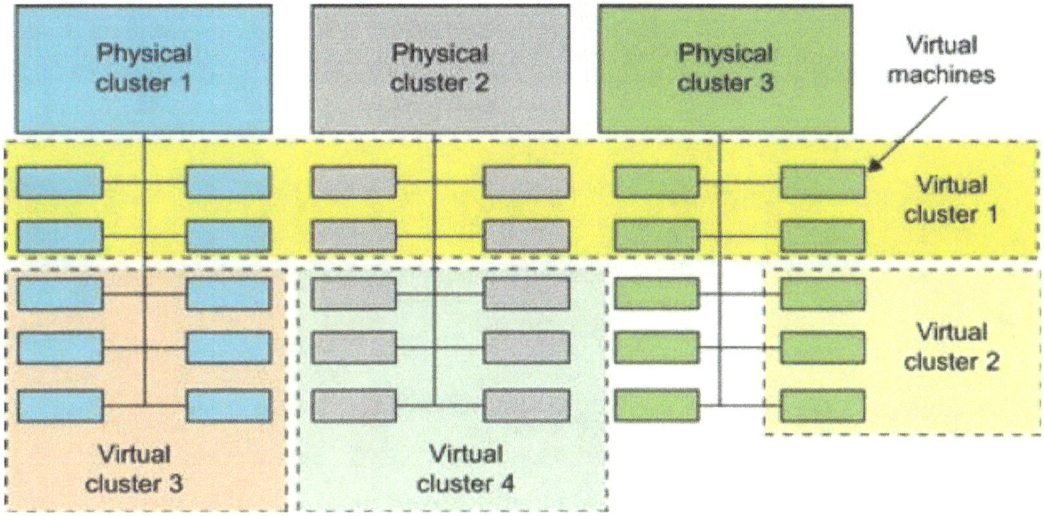

Figure 3.23 A cloud platform with four virtual clusters over three physical clusters shaded differently

Three physical clusters are shown on the left side of Figure 3.23. Four virtual clusters are created on the right, over the physical clusters. The physical machines are also called host systems. In contrast, the VMs are guest systems. The host and guest systems may run with different operating systems. Each VM can be installed on a remote server or replicated on multiple servers belonging to the same or different physical clusters. The boundary of a virtual cluster can change as VM nodes are added, removed, or migrated dynamically over time.

Figure 3.24 shows the concept of a virtual cluster based on application partitioning or customization. The different colors in the figure represent the nodes in different virtual clusters. As a large number of VM images might be present, the most important thing is to determine how to store those images in the system efficiently. There are common installations for most users or applications, such as operating systems or user-level programming libraries. These software packages can be preinstalled as templates (called template VMs). With these templates, users can build their own software stacks. New OS instances can be copied from the template VM. User-specific components such as programming libraries and applications can be installed to those instances.

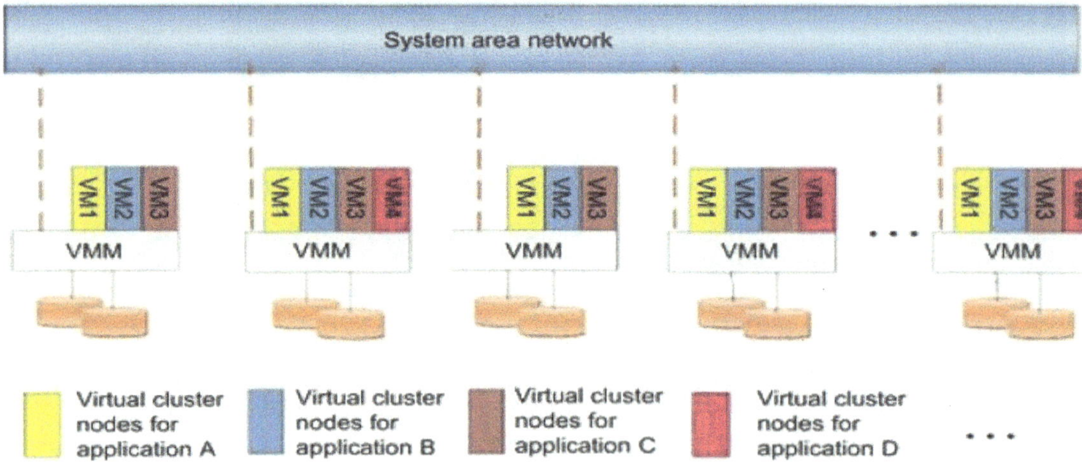

Figure 3.24 The concept of a virtual cluster based on application partitioning

3.7.1.1 Fast Deployment and Effective Scheduling

The system should have the capability of fast deployment. Deployment means two things:

- To construct and distribute software stacks (OS, libraries, applications) to a physical node inside clusters as fast as possible,
- To quickly switch runtime environments from one user's virtual cluster to another user's virtual cluster. If one user finishes using his system, the corresponding virtual cluster should shut down or suspend quickly to save the resources to run other VMs for other users.

Green computing

- Previous approaches have focused on saving the energy cost of components in a single workstation without a global vision. Consequently, they do not necessarily reduce the power consumption of the whole cluster.
- Other cluster-wide energy-efficient techniques can only be applied to homogeneous workstations and specific applications.
- The live migration of VMs allows workloads of one node to transfer to another node. However, it does not guarantee that VMs can randomly migrate among themselves.
- The challenge is to determine how to design migration strategies to implement green computing without influencing the performance of clusters.
- Another advantage of virtualization is load balancing of applications in a virtual cluster. Load balancing can be achieved using the load index and frequency of user logins.
- The automatic scale-up and scale-down mechanism of a virtual cluster can be implemented based on this model. Consequently, we can increase the resource utilization of nodes and shorten the response time of systems.

3.7.1.2 High-Performance Virtual Storage

The template VM can be distributed to several physical hosts in the cluster to customize the VMs. In addition, existing software packages reduce the time for customization as well as switching virtual environments.

It is important to efficiently manage the disk spaces occupied by template software packages.

Some storage architecture design can be applied to reduce duplicated blocks in a distributed file system of virtual clusters. Hash values are used to compare the contents of data blocks. Users have their own profiles which store the identification of the data blocks for corresponding VMs in a user-specific virtual cluster.

New blocks are created when users modify the corresponding data. Newly created blocks are identified in the users' profiles.

Basically, there are four steps to deploy a group of VMs onto a target cluster:

- preparing the disk image,
- configuring the VMs,
- choosing the destination nodes, and
- executing the VM deployment command on every host.

Many systems use templates to simplify the disk image preparation process.

A template is a disk image that includes a preinstalled operating system with or without certain application software. Templates could implement the COW (Copy on Write) format. A new COW backup file is very small and easy to create and transfer. Therefore, it definitely reduces disk space consumption.

Every VM is configured with a name, disk image, network setting, and allocated CPU and memory.

One needs to record each VM configuration into a file. However, this method is inefficient when managing a large group of VMs. VMs with the same configurations could use preedited profiles to simplify the process. In this scenario, the system configures the VMs according to the chosen profile.

3.7.2 Live VM Migration Steps and Performance Effects

Life migration

- In a cluster built with mixed nodes of host and guest systems, the normal method of operation is to run everything on the physical machine.
- When a VM fails, its role could be replaced by another VM on a different node, as long as they both run with the same guest OS.
- This is different from physical-to-physical failover in a traditional physical cluster. The advantage is enhanced failover flexibility.
- The potential drawback is that a VM must stop playing its role if its residing host node fails. However, this problem can be mitigated with VM life migration.

Figure 3.25 shows the process of life migration of a VM from host A to host B. The migration copies the VM state file from the storage area to the host machine.

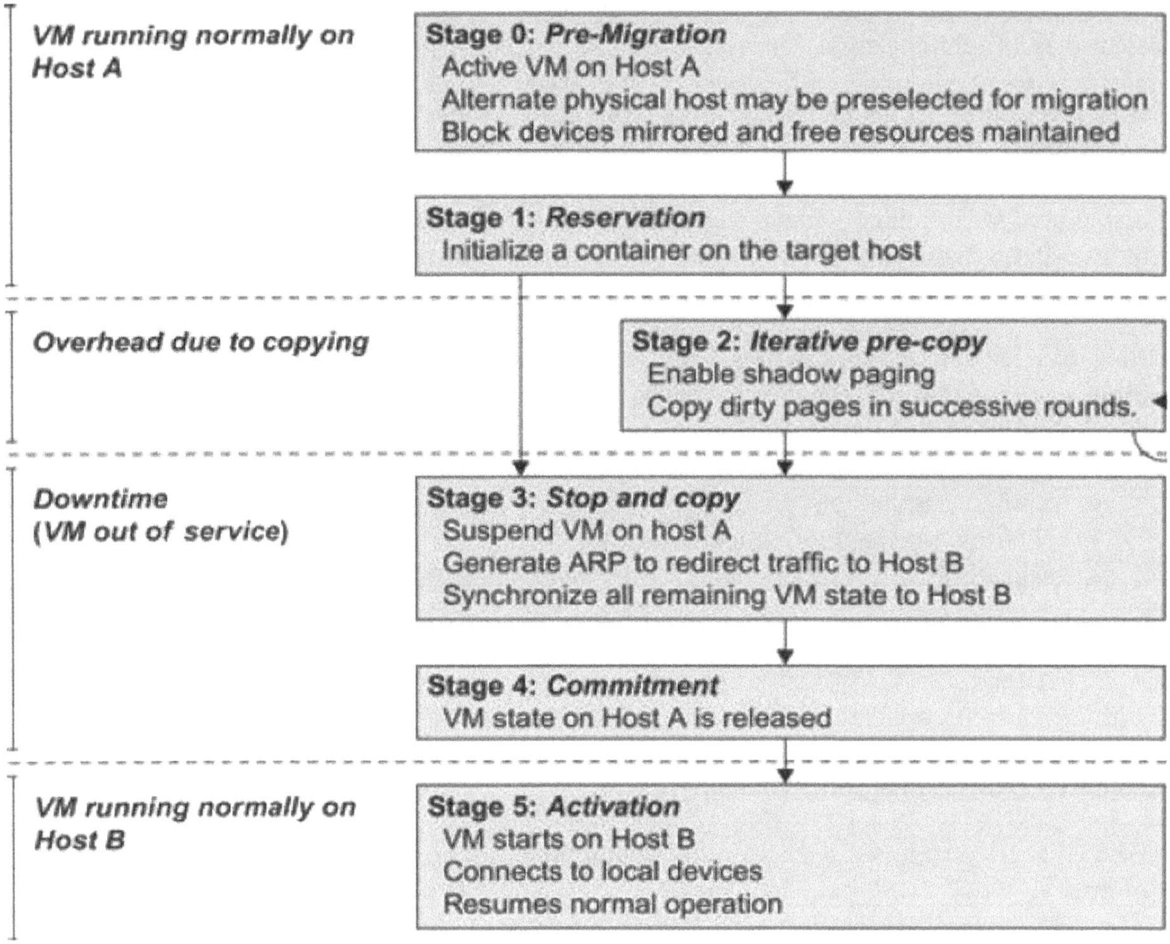

Figure 3.25 Live migration process of a VM from one host to another

There are four ways to manage a virtual cluster.

[1]. **First, you can use a guest-based manager, by which the cluster manager resides on a guest system. In this case, multiple VMs form a virtual cluster.**
For example, openMosix is an open source Linux cluster running different guest systems on top of the Xen hypervisor. Another example is Sun's cluster Oasis, an experimental Solaris cluster of VMs supported by a VMware VMM.

[2]. **Second, you can build a cluster manager on the host systems.**
The host-based manager supervises the guest systems and can restart the guest system on another physical machine. A good example is the VMware HA system that can restart a guest system after failure.

[3]. **A third way to manage a virtual cluster is to use an independent cluster manager on both the host and guest systems.**
This will make infrastructure management more complex,.

[4]. **finally, you can use an integrated cluster on the guest and host systems**. This means the manager must be designed to distinguish between virtualized resources and physical resources. Various cluster management schemes can be greatly enhanced when VM life migration is enabled with minimal overhead.

Furthermore, **we should ensure that the migration will not disrupt other active services residing** in the same host through resource contention (e.g., CPU, network bandwidth).

A VM can be in one of the following four states.

- An inactive state is defined by the virtualization platform, under which the VM is not enabled.
- An active state refers to a VM that has been instantiated at the virtualization platform to perform a real task.
- A paused state corresponds to a VM that has been instantiated but disabled to process a task or paused in a waiting state.
- A VM enters the suspended state if its machine file and virtual resources are stored back to the disk.

As shown in Figure 3.20, live migration of a VM consists of the following five steps:

Steps 0 and 1: Start migration. This step makes preparations for the migration, including determining the migrating VM and the destination host.

Steps 2: Transfer memory. Since the whole execution state of the VM is stored in memory, sending the VM's memory to the destination node ensures continuity of the service provided by the VM.

Step 3: Suspend the VM and copy the last portion of the data. The migrating VM's execution is suspended when the last round's memory data is transferred.

Steps 4 and 5: Commit and activate the new host. After all the needed data is copied, on the destination host, the VM reloads the states and recovers the execution of programs in it, and the service provided by this VM continues.

3.7.3 Migration of Memory, Files, and Network Resources

Since clusters have a high initial cost of ownership, including space, power conditioning, and cooling equipment, leasing or sharing access to a common cluster is an attractive solution when demands vary over time. Shared clusters offer economies of scale and more effective utilization of resources by multiplexing. Early configuration and management systems focus on expressive and scalable mechanisms for defining clusters for specific types of service, and physically partition cluster nodes among those types. When one system migrates to another physical node, we should consider the following issues.

3.7.3.1 Memory Migration

Moving the memory instance of a VM from one physical host to another is memory migration. The techniques employed for this purpose depend upon the characteristics of application/workloads supported by the guest OS.

Memory migration can be in a range of hundreds of megabytes to a few gigabytes in a typical system today, and it needs to be done in an efficient manner.

Technique for memory migration: Internet Suspend-Resume (ISR)

- Technique exploits temporal locality as memory states
- Temporal locality refers to the fact that the memory states differ only by the amount of work done since a VM was last suspended before being initiated for migration.
- To exploit temporal locality, each file in the file system is represented as a **tree of small subfiles**.
- A copy of this tree exists in both the suspended and resumed VM instances.
- The advantage of using a **tree-based representation** of files is that the caching ensures the transmission of only those files which have been changed.

The ISR technique deals with situations where the migration of **live machines is not a necessity**. Predictably, the downtime (the period during which the service is unavailable due to there being no currently executing instance of a VM) is high.

3.7.3.2 File System Migration

To support VM migration, a system must provide each VM with a consistent, location-independent view of the file system that is available on all hosts.

- A simple way to achieve this is to provide each VM with its own virtual disk which the file system is mapped to and transport the contents of this virtual disk along with the other states of the VM. However, due to the current trend of high-capacity disks, migration of the contents of an entire disk over a network is not a viable solution.
- Another way is to have a global file system across all machines where a VM could be located. This way removes the need to copy files from one machine to another because all files are network-accessible.

A **distributed file system** is used in ISR serving as a transport mechanism for propagating a suspended VM state.

- The actual file systems themselves are not mapped onto the distributed file system. Instead, the VMM only accesses its local file system.
- The relevant VM files are explicitly copied into the local file system for a resume operation and taken out of the local file system for a suspend operation.
- This approach relieves developers from the complexities of implementing several different file system calls for different distributed file systems.
- It also essentially disassociates the VMM from any particular distributed file system semantics. However, this decoupling means that the VMM has to store the contents of each VM's virtual disks in its local files, which have to be moved around with the other state information of that VM.

In **smart copying**, the VMM exploits spatial locality.

- Typically, people often move between the same small number of locations, such as their home and office. In these conditions, it is possible to transmit only the difference between the two file systems at suspending and resuming locations.

- This technique significantly reduces the amount of actual physical data that has to be moved. In situations where there is no locality to exploit, a different approach is to synthesize much of the state at the resuming site.

On many systems, user files only form a small fraction of the actual data on disk. Operating system and application software account for the majority of storage space. The proactive state transfer solution works in those cases where the resuming site can be predicted with reasonable confidence.

3.7.3.3 Network Migration

A migrating VM should maintain all open network connections without relying on forwarding mechanisms on the original host or on support from mobility or redirection mechanisms. To enable remote systems to locate and communicate with a VM, each VM must be assigned a virtual IP address known to other entities. This address can be distinct from the IP address of the host machine where the

VM is currently located. Each VM can also have its own distinct virtual MAC address. The VMM maintains a mapping of the virtual IP and MAC addresses to their corresponding VMs. In general, a migrating VM includes all the protocol states and carries its IP address with it.

Source and destination machines of a VM migration are typically connected to a single switched LAN, an unsolicited ARP reply from the migrating host is provided advertising that the IP has moved to a new location.

- This solves the open network connection problem by reconfiguring all the peers to send future packets to a new location. Although a few packets that have already been transmitted might be lost, there are no other problems with this mechanism. Alternatively, on a switched network, the migrating OS can keep its original Ethernet MAC address and rely on the network switch to detect its move to a new port.

Live migration means moving a VM from one physical node to another while keeping its OS environment and applications unbroken.

- This capability is being increasingly utilized in today's enterprise environments to provide efficient online system maintenance, reconfiguration, load balancing, and proactive fault tolerance. It provides desirable features to satisfy requirements for computing resources in modern computing systems, including server consolidation, performance isolation, and ease of management.

Cluster environment where a network-accessible storage system, such as storage area network (SAN) or network attached storage (NAS), is employed. Only memory and CPU status needs to be transferred from the source node to the target node.

Live migration techniques uses two approaches

- **Precopy approach**, which first transfers all memory pages, and then only copies modified pages during the last round iteratively. The VM service downtime is expected to be minimal by using iterative copy operations.
    - ➔ Performance degradation will occur because the **migration daemon** continually consumes network bandwidth to transfer dirty pages in each round. An **adaptive rate limiting approach** is employed to mitigate this issue.

- → Precopy approach are caused by the **large amount of transferred data** during the whole migration process. A checkpointing/recovery and trace/replay approach (CR/TRMotion) is proposed to provide fast VM migration.

- **Post copy approach,** all memory pages are transferred only once during the whole migration process and the baseline total migration time is reduced.
  - → The **downtime is much higher** than that of precopy due to the latency of fetching pages from the source node before the VM can be resumed on the target. With the advent of multicore or many-core machines, abundant CPU resources are available. Even if several VMs reside on a same multicore machine, CPU resources are still rich because physical CPUs are frequently amenable to multiplexing. We can exploit these copious CPU resources to compress page frames and the amount of transferred data can be significantly reduced. Memory compression algorithms typically have little memory overhead. Decompression is simple and very fast and requires no memory for decompression.

3.7.3.4 Live Migration of VM Using Xen

Xen hypervisor allows multiple commodity OSes to share x86 hardware in a safe and orderly fashion. The following example explains how to perform live migration of a VM between two Xen-enabled host machines. Domain 0 (or Dom0) performs tasks to create, terminate, or migrate to another host. Xen uses a send/recv model to transfer states across VMs.

Live Migration of VMs between Two Xen-Enabled Hosts

Xen supports live migration. It is a useful feature and natural extension to virtualization platforms that allows for the transfer of a VM from one physical machine to another with little or no downtime of the services hosted by the VM. Live migration transfers the working state and memory of a VM across a network when it is running. Xen also supports VM migration by using a mechanism called Remote Direct Memory Access (RDMA).

RDMA speeds up VM migration by avoiding TCP/IP stack processing overhead. RDMA implements a different transfer protocol whose origin and destination VM buffers must be registered before any transfer operations occur, reducing it to a "onesided" interface. Data communication over RDMA does not need to involve the CPU, caches, or context switches. This allows migration to be carried out with minimal impact on guest operating systems and hosted applications. Figure 3.26 shows the compression scheme for VM migration.

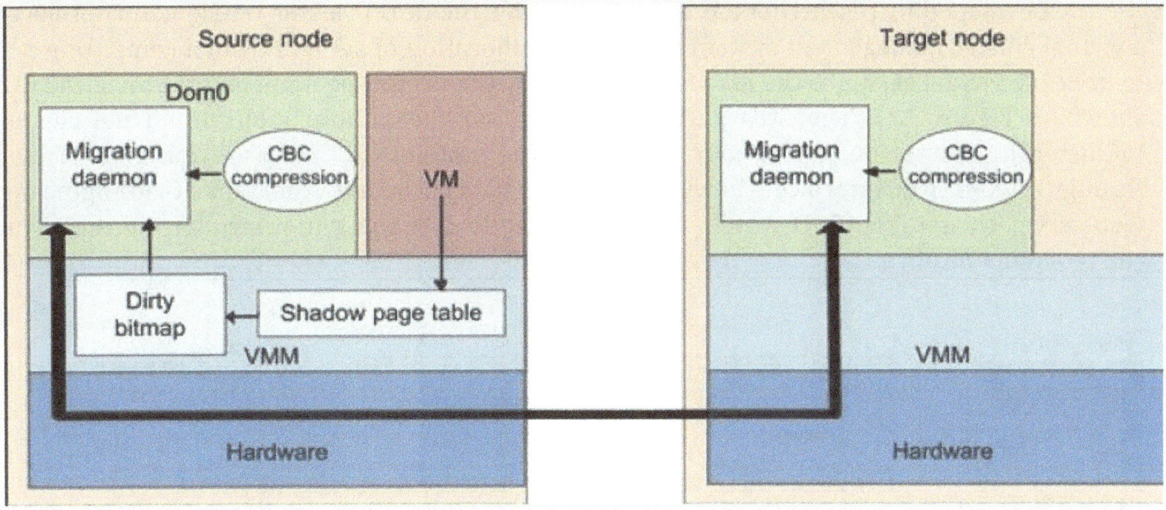

Figure 3.26 Live migration of VM from the Dom0 domain to a Xen-enabled target host.

Migration daemons running in the management VMs are responsible for performing migration. Shadow page tables in the VMM layer trace modifications to the memory page in migrated VMs during the precopy phase. Corresponding flags are set in a dirty bitmap. At the start of each precopy round, the bitmap is sent to the migration daemon. Then, the bitmap is cleared and the shadow page tables are destroyed and re-created in the next round. The system resides in Xen's management VM. Memory pages denoted by bitmap are extracted and compressed before they are sent to the destination. The compressed data is then decompressed on the target.

3.7.4 Dynamic Deployment of Virtual Clusters

Table 3.6 summarizes four virtual cluster research projects. The Cellular Disco at Stanford is a virtual cluster built in a shared-memory multiprocessor system. The INRIA virtual cluster was built to test parallel algorithm performance. The COD and VIOLIN clusters are studied in forthcoming examples.

Table 3.6 Experimental Results on Four Research Virtual Clusters

Project Name	Design Objectives	Reported Results and References
Cluster-on-Demand at Duke Univ.	Dynamic resource allocation with a virtual cluster management system	Sharing of VMs by multiple virtual clusters using Sun GridEngine
Cellular Disco at Stanford Univ.	To deploy a virtual cluster on a shared-memory multiprocessor	VMs deployed on multiple processors under a VMM called Cellular Disco
VIOLIN at Purdue Univ	Multiple VM clustering to prove the advantage of dynamic adaptation	Reduce execution time of applications running VIOLIN with adaptation
GRAAL Project at INRIA in France	Performance of parallel algorithms in Xen enabled virtual clusters	75% of max. performance achieved with 30% resource slacks over VM clusters

Example: The Cluster-on-Demand (COD) Project at Duke University

Developed by researchers at Duke University, the COD (Cluster-on-Demand) project is a virtual cluster management system for dynamic allocation of servers from a computing pool to multiple virtual clusters. The idea is illustrated by the prototype implementation of the COD shown in Figure 3.27. The COD partitions a physical cluster into multiple virtual clusters (vClusters). vCluster owners specify the operating systems and software for their clusters through an XML-RPC interface. The vClusters run a batch schedule from Sun's GridEngine on a web server cluster. The COD system can respond to load changes in restructuring the virtual clusters dynamically.

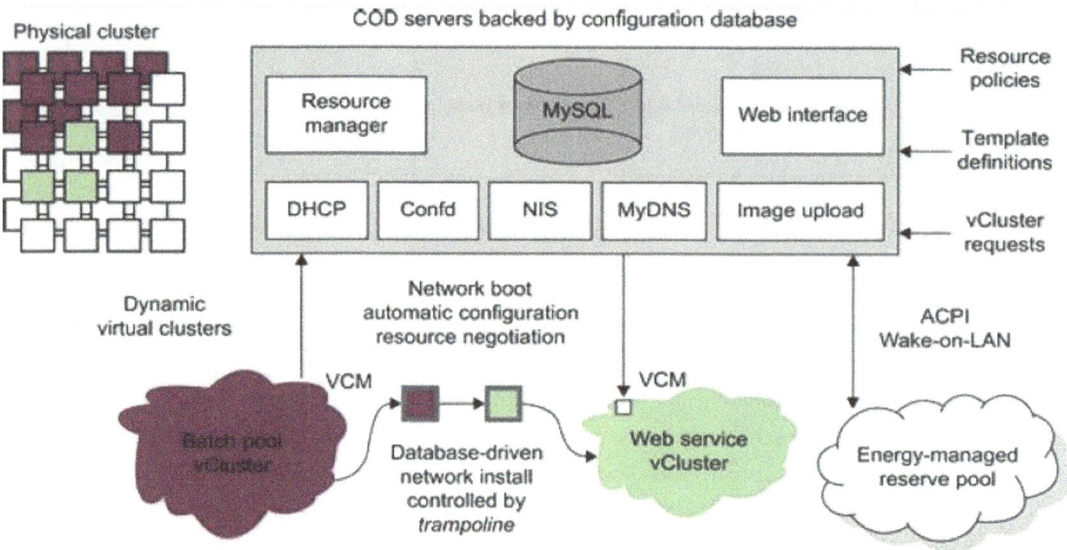

Figure 3.27 COD partitioning a physical cluster into multiple virtual clusters

Example: The VIOLIN Project at Purdue University

The Purdue VIOLIN Project applies live VM migration to reconfigure a virtual cluster environment. Its purpose is to achieve better resource utilization in executing multiple cluster jobs on multiple cluster domains. The project leverages the maturity of VM migration and environment adaptation technology. The approach is to enable mutually isolated virtual environments for executing parallel applications on top of a shared physical infrastructure consisting of multiple domains. Figure 3.28 illustrates the idea with five concurrent virtual environments, labeled as VIOLIN 1–5, sharing two physical clusters.

The squares of various shadings represent the VMs deployed in the physical server nodes. The major contribution by the Purdue group is to achieve autonomic adaptation of the virtual computation environments as active, integrated entities. A virtual execution environment is able to relocate itself across the infrastructure, and can scale its share of infrastructural resources. The adaptation is transparent to both users of virtual environments and administrations of infrastructures. The adaptation overhead is maintained at 20 sec out of 1,200 sec in solving a large NEMO3D problem of 1 million particles.

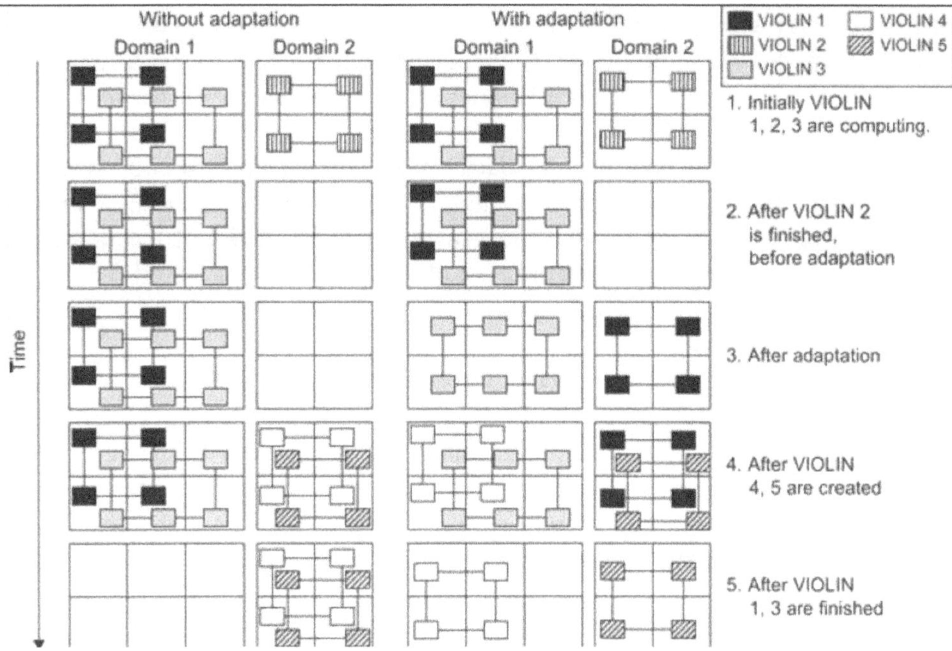

Figure 3.28 VIOLIN adaptation scenarios of five virtual environments sharing two hosted clusters. Note that there are more idle squares (blank nodes) before and after the adaptation.

3.8 Virtualization for Data-Center Automation

Data centers have grown rapidly in recent years, and all major IT companies are pouring their resources into building new data centers. In addition, Google, Yahoo!, Amazon, Microsoft, HP, Apple, and IBM are all in the game. All these companies have invested billions of dollars in datacenter construction and automation. Data-center automation means that huge volumes of hardware, software, and database resources in these data centers can be allocated dynamically to millions of Internet users simultaneously, with guaranteed QoS and cost-effectiveness.

This automation process is triggered by the growth of virtualization products and cloud computing services. The latest virtualization development highlights high availability (HA), backup services, workload balancing, and further increases in client bases. IDC projected that automation, service orientation, policy-based, and variable costs in the virtualization market.

3.8.1 Server Consolidation in Data Centers

In data centers, a large number of heterogeneous workloads can run on servers at various times. These heterogeneous workloads can be roughly divided into two categories:

- **Chatty workloads** - these may burst at some point and return to a silent state at some other point. A web video service is an example of this, whereby a lot of people use it at night and few people use it during the day.

- **Noninteractive workloads**- these workloads do not require people's efforts to make progress after they are submitted. High-performance computing is a typical example of this. At various stages, the requirements for resources of these workloads are dramatically different.

However, to guarantee that a workload will always be able to cope with all demand levels, the workload is statically allocated enough resources so that peak demand is satisfied. Therefore, it is common that most servers in **data centers are underutilized.** A large amount of hardware, space, power, and management cost of these servers is wasted.

- **Server consolidation is an approach** to improve the low utility ratio of hardware resources by reducing the number of physical servers. Among several server consolidation techniques such as centralized and physical consolidation, virtualization-based server consolidation is the most powerful. Data centers need to optimize their resource management. Yet these techniques are performed with the granularity of a full server machine, which makes resource management far from well optimized. Server virtualization enables smaller resource allocation than a physical machine

The use of VMs increases resource management complexity. This causes a challenge in terms of how to improve resource utilization as well as guarantee QoS in data centers. In detail, server virtualization has the following side effects:

- Consolidation enhances hardware utilization. Many underutilized servers are consolidated into fewer servers to enhance resource utilization. Consolidation also facilitates backup services and disaster recovery.
- This approach enables more agile provisioning and deployment of resources. In a virtual environment, the images of the guest OSes and their applications are readily cloned and reused.
- The total cost of ownership is reduced. In this sense, server virtualization causes deferred purchases of new servers, a smaller data-center footprint, lower maintenance costs, and lower power, cooling, and cabling requirements.
- This approach improves availability and business continuity. The crash of a guest OS has no effect on the host OS or any other guest OS. It becomes easier to transfer a VM from one server to another, because virtual servers are unaware of the underlying hardware.

To automate data-center operations, one must consider

- resource scheduling
- architectural support
- power management
- automatic or autonomic resource management
- performance of analytical models, and so on.

In **virtualized data centers**, an efficient, on-demand, fine-grained scheduler is one of the key factors to improve resource utilization. Scheduling and reallocations can be done in a wide range of levels in a set of data centers.

- The levels match at least at the VM level, server level, and data-center level. Ideally, scheduling and resource reallocations should be done at all levels. However, due to the complexity of this, current techniques only focus on a single level or, at most, two levels.

Dynamic CPU allocation is based on VM utilization and application-level QoS metrics.

- One method considers both CPU and memory flowing as well as automatically adjusting resource overhead based on varying workloads in hosted services.
- Another scheme uses a two-level resource management system to handle the complexity involved. A local controller at the VM level and a global controller at the server level are designed. They implement autonomic resource allocation via the interaction of the local and global controllers. Multicore and virtualization are two cutting techniques that can enhance each other.

3.8.2 Virtual Storage Management

The term "storage virtualization" was widely used before the renaissance of system virtualization. Storage virtualization was largely used to describe the aggregation and repartitioning of disks at very coarse time scales for use by physical machines. Generally, the data stored can be classified into two categories: VM images and application data. The VM images are special to the virtual environment, while application data includes all other data which is the same as the data in traditional OS environments.

Encapsulation and isolation - Traditional operating systems and applications running on them can be encapsulated in VMs. Only one operating system runs in a virtualization while many applications run in the operating system. System virtualization allows multiple VMs to run on a physical machine and the VMs are completely isolated. To achieve encapsulation and isolation, both the system software and the hardware platform, such as CPUs and chipsets, are rapidly updated.

Storage is lagging - The storage systems become the main bottleneck of VM deployment. In virtualization environments, a virtualization layer is inserted between the hardware and traditional operating systems or a traditional operating system is modified to support virtualization. This procedure complicates storage operations.

VMs are not nimble - Hence, operations such as remapping volumes across hosts and check pointing disks are frequently clumsy and esoteric, and sometimes simply unavailable. Since traditional storage management techniques do not consider the features of storage in virtualization environments, Parallax designs a novel architecture in which storage features that have traditionally been implemented directly on high-end storage arrays and switchers are relocated into a federation of storage VMs. These storage VMs share the same physical hosts as the VMs that they serve.

Example: Parallax Providing Virtual Disks to Client VMs from a Large Com m on Share d Physical Disk

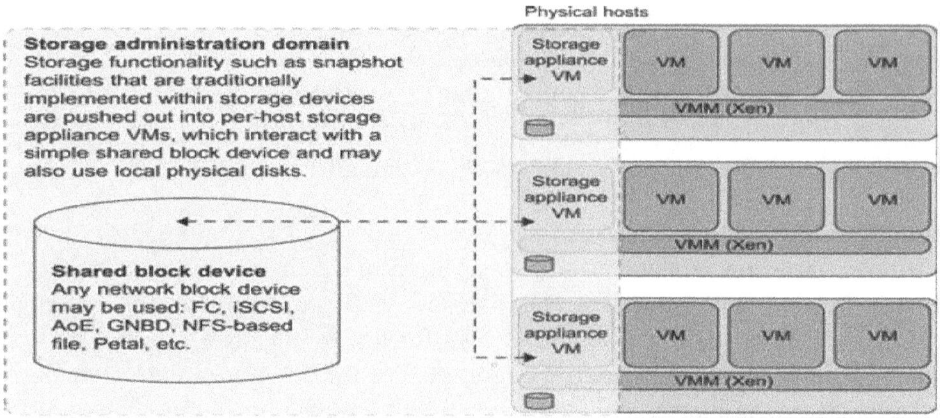

Figure 3.29 Parallax is a set of per-host storage appliances that share access to a common block device and presents virtual disks to client VMs.

Parallax itself runs as a user-level application in the storage appliance VM. It provides virtual disk images (VDIs) to VMs. A VDI is a single-writer virtual disk which may be accessed in a location-transparent manner from any of the physical hosts in the Parallax cluster. The VDIs are the core abstraction provided by Parallax. Parallax uses Xen's block tap driver to handle block requests and it is implemented as a tapdisk library. This library acts as a single block virtualization service for all client VMs on the same physical host. In the Parallax system, it is the storage appliance VM that connects the physical hardware device for blocks and network access. As shown in Figure 3.29, physical device drivers are included in the storage appliance VM. This implementation enables a storage administrator to live-upgrade the block device drivers in an active cluster.

3.8.3 Cloud OS for Virtualized Data Centers

Data centers must be virtualized to serve as cloud providers. Table 3.7 summarizes four virtual infrastructure (VI) managers and OSes. These VI managers and OSes are specially tailored for virtualizing data centers which often own a large number of servers in clusters. Nimbus, Eucalyptus, and OpenNebula are all open source software available to the general public. Only vSphere 4 is a proprietary OS for cloud resource virtualization and management over data centers.

These VI managers are used to create VMs and aggregate them into virtual clusters as elastic resources. Nimbus and Eucalyptus support essentially virtual networks. OpenNebula has additional features to provision dynamic resources and make advance reservations. All three public VI managers apply Xen and KVM for virtualization. vSphere 4 uses the hypervisors ESX and ESXi from VMware. Only vSphere 4 supports virtual storage in addition to virtual networking and data protection. We will study Eucalyptus and vSphere 4 in the next two examples.

Table 3.7 VI Managers and Operating Systems for Virtualizing Data Centers

Manager/ OS, Platforms, License	Resources Being Virtualized, Web Link	Client API, Language	Hypervisors Used	Public Cloud Interface	Special Features
Nimbus Linux, Apache v2	VM creation, virtual cluster, www .nimbusproject.org/	EC2 WS, WSRF, CLI	Xen, KVM	EC2	Virtual networks
Eucalyptus Linux, BSD	Virtual networking (Example 3.12 and [41]), www .eucalyptus.com/	EC2 WS, CLI	Xen, KVM	EC2	Virtual networks
OpenNebula Linux, Apache v2	Management of VM, host, virtual network, and scheduling tools, www.opennebula.org/	XML-RPC, CLI, Java	Xen, KVM	EC2, Elastic Host	Virtual networks, dynamic provisioning
vSphere 4 Linux, Windows, proprietary	Virtualizing OS for data centers (Example 3.13), www .vmware.com/ products/vsphere/ [66]	CLI, GUI, Portal, WS	VMware ESX, ESXi	VMware vCloud partners	Data protection, vStorage, VMFS, DRM, HA

Example: Eucalyptus for Virtual Networking of Private Cloud

Eucalyptus is an open source software system (Figure 3.30) intended mainly for supporting Infrastructure as a Service (IaaS) clouds. The system primarily supports virtual networking and the management of VMs; virtual storage is not supported. Its purpose is to build private clouds that can interact with end users through Ethernet or the Internet. The system also supports interaction with other private clouds or public clouds over the Internet. The system is short on security and other desired features for general purpose grid or cloud applications.

The three resource managers in Figure 3.30 are specified below:

- Instance Manager controls the execution, inspection, and terminating of VM instances on the host where it runs.
- Group Manager gathers information about and schedules VM execution on specific instance managers, as well as manages virtual instance network.
- Cloud Manager is the entry-point into the cloud for users and administrators. It queries node managers for information about resources, makes scheduling decisions, and implements them by making requests to group managers.

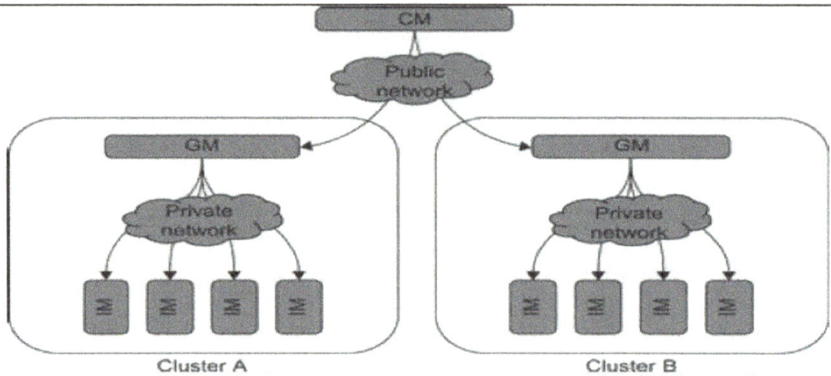

Figure 3.30 Eucalyptus for building private clouds by establishing virtual networks over the VMs linking through Ethernet and the Internet

Example: VMware vSphere4 as a Commercial Cloud OS

The vSphere 4 offers a hardware and software ecosystem developed by VMware and released in April 2009. vSphere extends earlier virtualization software products by VMware, namely the VMware Workstation, ESX for server virtualization, and Virtual Infrastructure for server clusters. Figure 3.31 shows vSphere's overall architecture. The system interacts with user applications via an interface layer, called vCenter. vSphere is primarily intended to offer virtualization support and resource management of datacenter resources in building private clouds. VMware claims the system is the first cloud OS that supports availability, security, and scalability in providing cloud computing services.

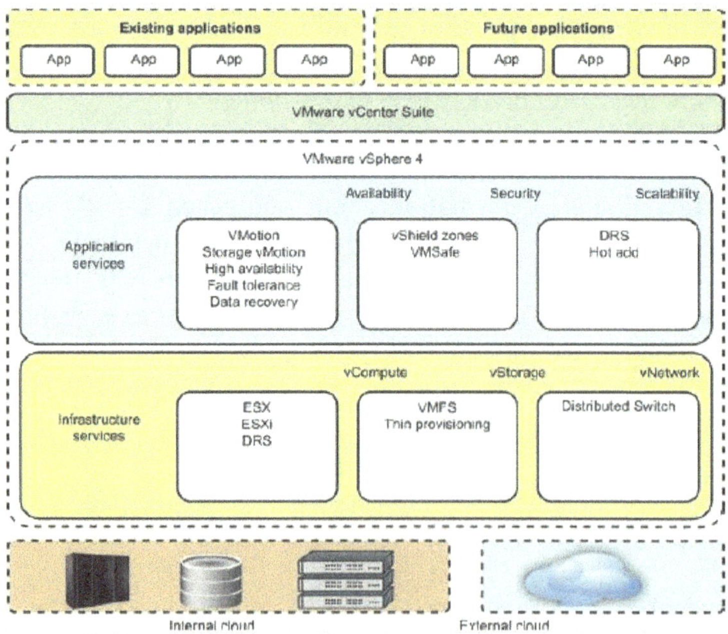

Figure 3.31 vSphere/4, a cloud operating system that manages compute, storage, and network resources over virtualized data centers.

The vSphere 4 is built with two functional software suites: infrastructure services and application services. It also has three component packages intended mainly for virtualization purposes: vCompute is supported by ESX, ESXi, and DRS virtualization libraries from VMware; vStorage is supported by VMS and thin provisioning libraries; and vNetwork offers distributed switching and networking functions. These packages interact with the hardware servers, disks, and networks in the data center. These infrastructure functions also communicate with other external clouds.

The application services are also divided into three groups: availability, security, and scalability. Availability support includes VMotion, Storage VMotion, HA, Fault Tolerance, and Data Recovery from VMware. The security package supports vShield Zones and VMsafe. The scalability package was built with DRS and Hot Add. Interested readers should refer to the vSphere 4 web site for more details regarding these component software functions. To fully understand the use of vSphere 4, users must also learn how to use the vCenter interfaces in order to link with existing applications or to develop new applications.

3.8.4 Trust Management in Virtualized Data Centers

A VM entirely encapsulates the state of the guest operating system running inside it. Encapsulated machine state can be copied and shared over the network and removed like a normal file, which proposes a challenge to VM security. In general, a VMM can provide secure isolation and a VM accesses hardware resources through the control of the VMM, so the VMM is the base of the security of a virtual system. Normally, one VM is taken as a management VM to have some privileges such as creating, suspending, resuming, or deleting a VM.

Once a hacker successfully enters the VMM or management VM, the whole system is in danger. A subtler problem arises in protocols that rely on the **"freshness"** of their random number source for generating session keys. Considering a VM, rolling back to a point after a random number has been chosen, but before it has been used, resumes execution; the random number, which must be "fresh" for security purposes, is reused. With a stream cipher, two different plaintexts could be encrypted under the same key stream, which could, in turn, expose both plaintexts if the plaintexts have sufficient redundancy. Non cryptographic protocols that rely on freshness are also at risk. For example, the reuse of TCP initial sequence numbers can raise TCP hijacking attacks.

3.8.4.1 VM-Based Intrusion Detection

Intrusions are unauthorized access to a certain computer from local or network users and intrusion detection is used to recognize the unauthorized access. An intrusion detection system (IDS) is built on operating systems, and is based on the characteristics of intrusion actions. Typical IDS can be classified as **host-based IDS (HIDS)** or **network-based IDS (NIDS),** depending on the data source.

A HIDS can be implemented on the monitored system. When the monitored system is attacked by hackers, the HIDS also faces the risk of being attacked. A NIDS is based on the flow of network traffic which can't detect fake actions.

Virtualization-based intrusion detection can isolate guest VMs on the same hardware platform. Even some VMs can be invaded successfully; they never influence other VMs, which is similar to the way in which a NIDS operates. Furthermore, a VMM monitors and audits access requests for hardware and system software. This can avoid fake actions and possess the merit of a HIDS. There are two different methods for implementing a VM-based IDS: Either the IDS is an independent process in each VM or a high-privileged VM on the VMM; or the IDS is integrated into the VMM and has the same privilege to access the hardware as well as the VMM. Figure 3.32 illustrates the concept.

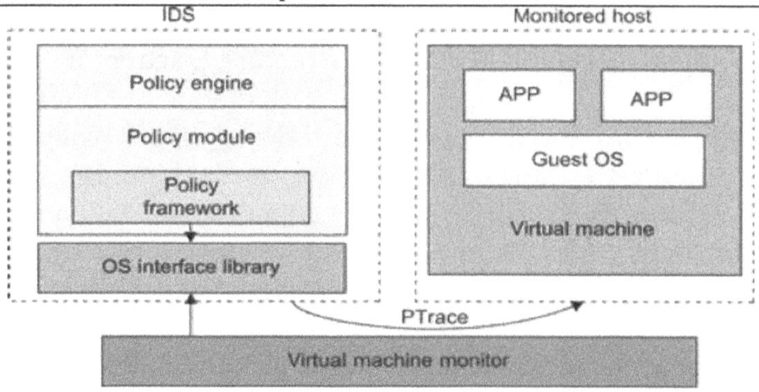

Figure 3.32 The architecture of livewire for intrusion detection using a dedicated VM

The VM-based IDS contains a **policy engine** and a **policy module**.

- The policy framework can monitor events in different guest VMs by operating system interface library and PTrace indicates trace to secure policy of monitored host.
- It's difficult to predict and prevent all intrusions without delay. Therefore, an analysis of the intrusion action is extremely important after an intrusion occurs.

The **IDS log service** is based on the operating system kernel. Thus, when an operating system is invaded by attackers, the log service should be unaffected.

Besides IDS, **honeypots and honeynets** are also prevalent in intrusion detection. They attract and provide a fake system view to attackers in order to protect the real system. In addition, the attack action can be analyzed, and a secure IDS can be built.

A **honeypot** is a purposely defective system that simulates an operating system to cheat and monitor the actions of an attacker. A honeypot can be divided into physical and virtual forms. A guest operating system and the applications running on it constitute a VM. The host operating system and VMM must be guaranteed to prevent attacks from the VM in a virtual honeypot.

Example: EMC Establishment of Trusted Zones for Protection of Virtual Clusters Provided to Multiple Tenants

EMC and VMware have joined forces in building security middleware for trust management in distributed systems and private clouds. The concept of trusted zones was established as part of the virtual infrastructure. Figure 3.33 illustrates the concept of creating trusted zones for virtual clusters (multiple applications and OSes for each tenant) provisioned in separate virtual environments. The physical infrastructure is shown at the bottom, and marked as a cloud provider. The virtual clusters or infrastructures are shown in the upper boxes for two tenants. The public cloud is associated with the global user communities at the top.

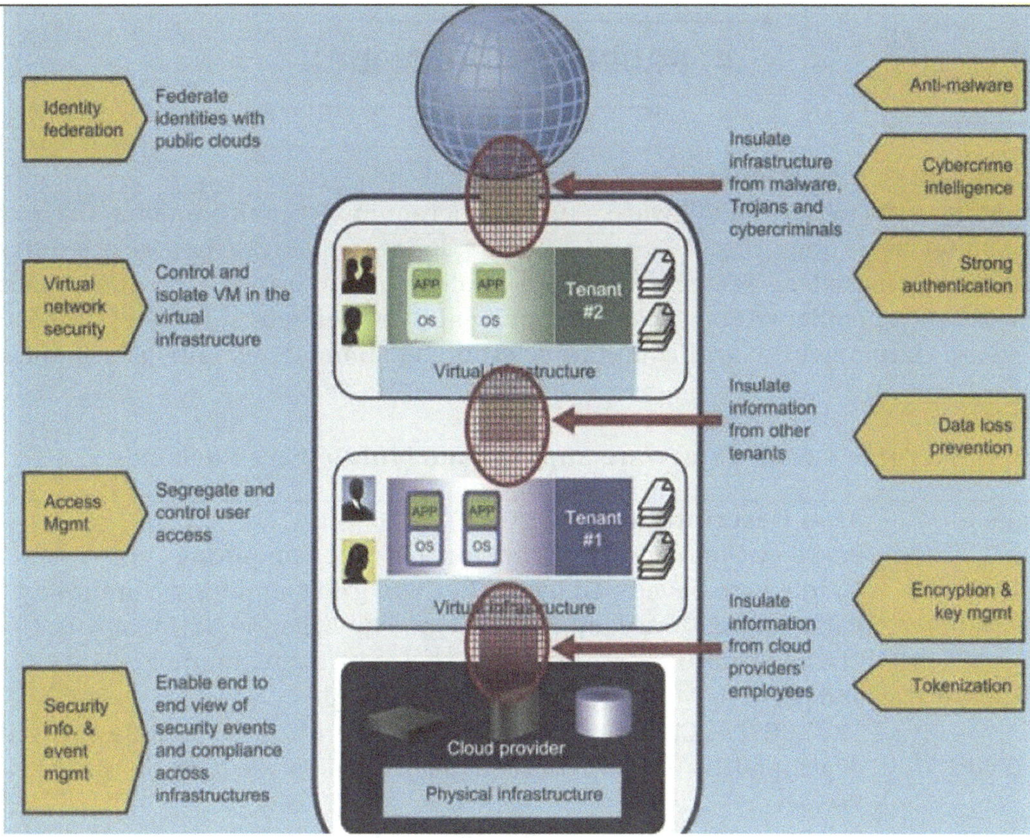

Figure 3.33 Techniques for establishing trusted zones for virtual cluster insulation and VM isolation

The arrowed boxes on the left and the brief description between the arrows and the zoning boxes are security functions and actions taken at the four levels from the users to the providers. The small circles between the four boxes refer to interactions between users and providers and among the users themselves. The arrowed boxes on the right are those functions and actions applied between the tenant environments, the provider, and the global communities.

Almost all available countermeasures, such as anti-virus, worm containment, intrusion detection, encryption and decryption mechanisms, are applied here to insulate the trusted zones and isolate the VMs for private tenants. The main innovation here is to establish the trust zones among the virtual clusters. The end result is to enable an end-to-end view of security events and compliance across the virtual clusters dedicated to different tenants.

4. PROGRAMMING MODEL

4.1 Open Source Grid Middleware Packages

Many software, middleware, and programming environments have been developed for grid computing over past 15 years. Below we assess their relative strength and limitations based on recently reported applications. We first introduce some grid standards and popular APIs. Then we present the desired software support and middleware developed for grid computing. Table 7.6 summarizes four grid middleware packages.

Table 4.1 Grid Software Support and Middleware Packages

Package	Brief Description
BOINC	Berkeley Open Infrastructure for Network Computing
UNICORE	Middleware developed by the German grid computing community
Globus (GT4) A	middleware library jointly developed by Argonne National Lab., Univ. of Chicago, and USC Information Science Institute, funded by DARPA, NSF, and NIH.
CGSP in ChinaGrid	The CGSP (ChinaGrid Support Platform) is a middleware library developed by 20 top universities in China as part of the ChinaGrid Project
Condor-G	Originally developed at the Univ. of Wisconsin for general distributed computing, and later extended to Condor-G for grid job management
Sun Grid Engine (SGE)	Developed by Sun Microsystems for business grid applications. Applied to private grids and local clusters within enterprises or campuses

4.1.1 Grid Standards and APIs

Two well-formed standards

- The Open Grid Forum (formally Global Grid Forum) and
- Object Management Group

some grid standards including

- GLUE for resource representation
- SAGA (Simple API for Grid Applications)
- GSI (Grid Security Infrastructure)
- OGSI (Open Grid Service Infrastructure)
- WSRE (Web Service Resource Framework)

The grid standards have guided the development of several middleware libraries and API tools for grid computing.

Research grids tested include the EGEE, France Grilles, D-Grid (German), CNGrid (China), TeraGrid (USA), etc. Production grids built with the standards include the EGEE, INFN grid (Italian), NorduGrid, Sun Grid, Techila, and Xgrid.

4.1.2 Software Support and Middleware

- Grid middleware is specifically designed a layer between hardware and the software.
- The middleware products enable the sharing of heterogeneous resources and managing virtual organizations created around the grid. Middleware glues the allocated resources with specific user applications.
- Popular grid middleware tools include the
    - Globus Toolkits (USA), gLight,
    - UNICORE (German),
    - BOINC (Berkeley),
    - CGSP (China),
    - Condor-G,
    - Sun Grid Engine, etc.

The following first example introduces the matchmaking capability of Condor and Condor-G developed at University of Wisconsin. Example second discusses the features of Sun's Grid Engine developed at Sun Microsystems for business grid computing.

Example: Features in Condor Kernel and Condor-G for Grid Computing

Condor is a software tool for high-throughput distributed batch computing.

- It was designed to explore the idle cycles of a network of distributed computers.
- The major components of Condor are the user agent, resources, and matchmaker, as shown in Figure 4.1.
- The ClassAds (classified advertisements) language was exploited in Condor to express user requests against available resources in a distributed system.
- Agents and resources advertise their status and requirements in ClassAds to a central matchmaker.
- The matchmaker scans the ClassAds and creates pairs of (resources, agents) that match each other's requirements.

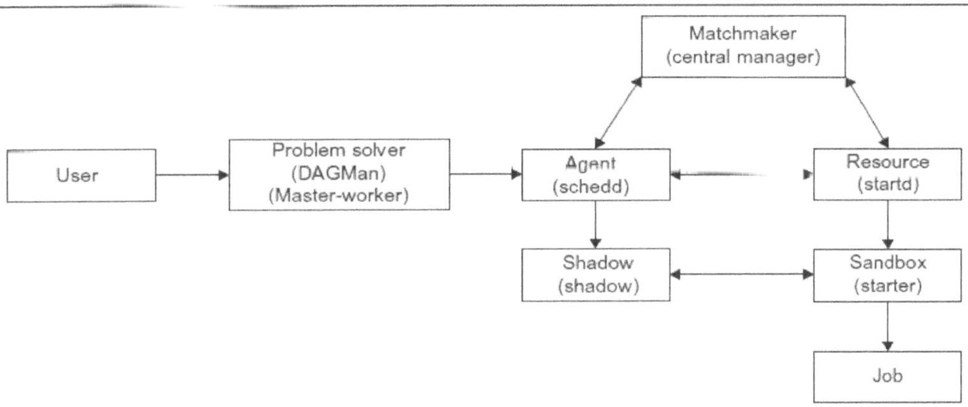

Figure 4.1 Major functional components in a Condor system (Condor-specific names are in parentheses)

Subsequently, the matched agent negotiates with the available resource to execute the job. Two problem solvers are provided in Condor:

- the master-worker
- the DAG manager

For specific jobs, Condor records checkpoints and subsequently resumes program execution from the checkpoints. When running a job on remote machines without a shared file system, Condor accesses the local execution environment via remote system calls. Condor-G was designed to match the GRAM protocol for coupling jobs with resources in the Globus Project. Condor-G adds durability and two-phase commitment to prevent the loss and repetition of jobs in GRAM. However, Condor-G does not support all the features in GRAM.

Example: The Sun Grid Engine (SGE) Middleware Package

SGE was developed by Sun Microsystems in response to increasing demands of business grid applications. SGE features are applied to private grids and local clusters within an enterprise or campus for intranet-based cluster or grid applications. The system supports batch processing with dynamic allocation of grid resources. Fault tolerance and failover capability are implemented. Users can specify resources independent of submission locations. Job status and resource utilization can be monitored periodically.

Sun offers free and enterprise editions of the SGE package. The system offers centralized management of all resources allocated to individual jobs. This was meant to enhance efficiency and performance. Suspend and resume tools are provided to allow users to halt a job and restart it later without losing the work already completed. The SGE workflow follows the following sequence of events:

- Accepts the jobs from users
- Places jobs in a computer area until execution
- Sends the jobs from the holding area to a host where they can be executed
- Manages the jobs during execution
- Logs a record of their execution when they are finished

The SGE system uses reserved ports, Kerberos, DCE, SSL, and authentication in classified hosts to enforce security at different trust levels and resource access restrictions.

4.2 The Globus Toolkit Architecture (GT4)

The Globus Toolkit, started in 1995 with funding from DARPA, is an open middleware library for the grid computing communities. These open source software libraries support many operational grids and their applications on an international basis. The toolkit addresses common problems and issues related to grid resource discovery, management, communication, security, fault detection, and portability. The software itself provides a variety of components and capabilities. The library includes a rich set of service implementations.

The implemented software supports grid infrastructure management, provides tools for building new web services in Java, C, and Python, builds a powerful standard-based security infrastructure and client APIs (in different languages), and offers comprehensive command-line programs for accessing various grid services. The Globus Toolkit was initially motivated by a desire to remove obstacles that prevent seamless collaboration, and thus sharing of resources and services, in scientific and engineering applications. The shared resources can be computers, storage, data, services, networks, science instruments (e.g., sensors), and so on. The Globus library version GT4, is conceptually shown in Figure 4.2.

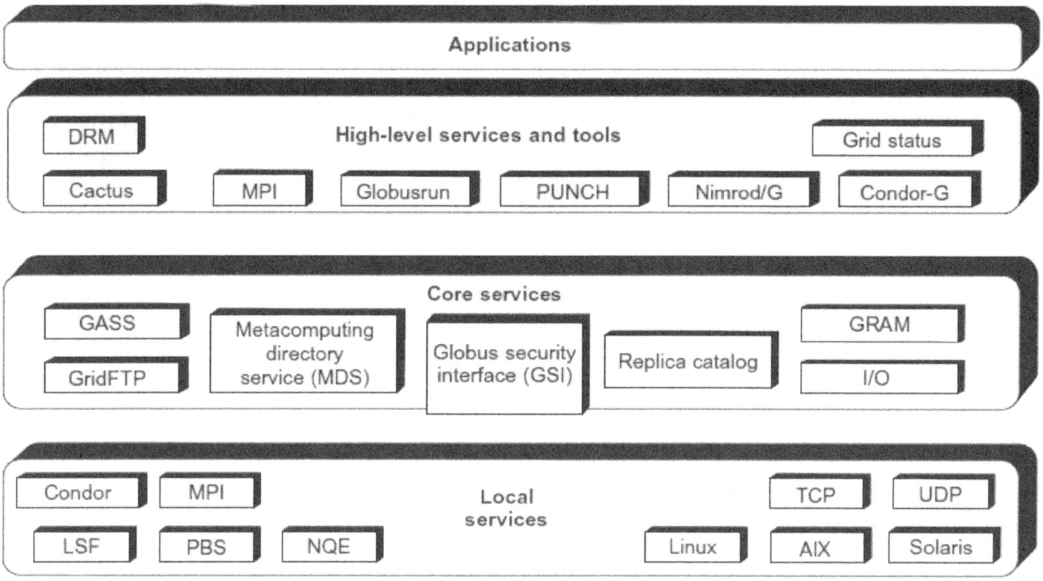

Figure 4.2 Globus Tookit GT4 supports distributed and cluster computing services.

4.2.1 The GT4 Library

GT4 offers the middle-level core services in grid applications.

- The **high-level services and tools**, such as MPI, Condor-G, and Nirod/G, are developed by third parties for general-purpose distributed computing applications.
- The **local services**, such as LSF, TCP, Linux, and Condor, are at the bottom level and are fundamental tools supplied by other developers.
- Table 4.2 summarizes GT4's core grid services by module name.
- Nexus is used for collective communications and HBM for heartbeat monitoring of resource nodes. GridFTP is for speeding up inter node file transfers. The module GASS is used for global access of secondary storage.

Table 4.2 Functional Modules in Globus GT4 Library

Service Functionality	Module Name	Functional Description
Global Resource Allocation Manager	GRAM	Grid Resource Access and Management (HTTP-based)
Communication	Nexus	Unicast and multicast communication
Grid Security Infrastructure	GSI	Authentication and related security services
Monitory and Discovery Service	MDS	Distributed access to structure and state information
Health and Status	HBM	Heartbeat monitoring of system components
Global Access of Secondary Storage	GASS	Grid access of data in remote secondary storage
Grid File Transfer	GridFTP	Inter-node fast file transfer

4.2.2 Globus Job Workflow

Figure 4.3 shows the typical job workflow when using the Globus tools. A typical job execution sequence proceeds as follows:

- The user delegates his credentials to a delegation service.
- The user submits a job request to GRAM with the delegation identifier as a parameter.
  - GRAM parses the request, retrieves the user proxy certificate from the delegation service, and then acts on behalf of the user.
  - GRAM sends a transfer request to the RFT (Reliable File Transfer), which applies GridFTP to bring in the necessary files.
  - GRAM invokes a local scheduler via a GRAM adapter and the SEG (Scheduler Event Generator) initiates a set of user jobs.
- The local scheduler reports the job state to the SEG. Once the job is complete, GRAM uses RFT and GridFTP to stage out the resultant files.
- The grid monitors the progress of these operations and sends the user a notification when they succeed, fail, or are delayed.

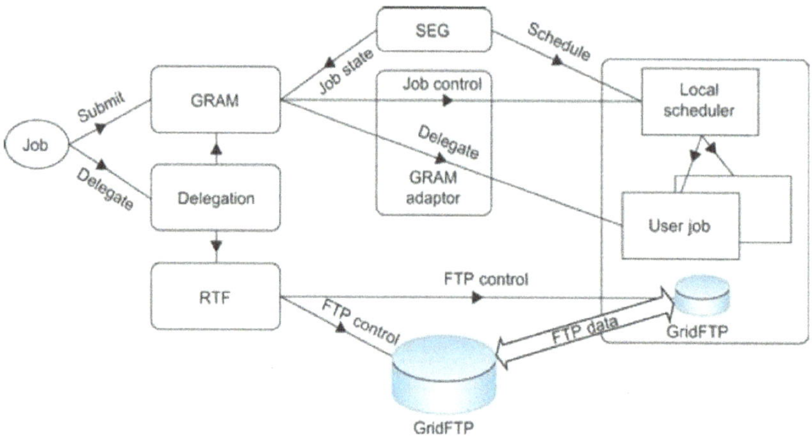

Figure 4.3 Globus job workflow among interactive functional modules

4.2.3 Client-Globus Interactions

GT4 service programs are designed to support user applications as illustrated in Figure 4.4.

- There are strong interactions between provider programs and user code.
- GT4 makes heavy use of industry standard web service protocols and mechanisms in service description, discovery, access, authentication, authorization, and the like. GT4 makes extensive use of Java, C, and Python to write user code.
- Web service mechanisms define specific interfaces for grid computing. Web services provide flexible, extensible, and widely adopted XML-based interfaces.

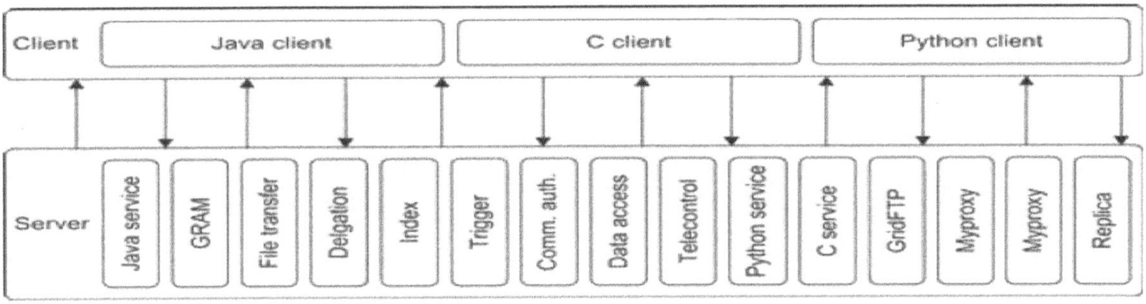

Figure 4.4 Client and GT4 server interactions; vertical boxes correspond to service programs and horizontal boxes represent the user codes.

- GT4 provides a set of infrastructure services for accessing, monitoring, managing, and controlling access to infrastructure elements.
- GT4 implements standards to facilitate construction of operable and reusable user code.
- Developers can use these services and libraries to build simple and complex systems quickly.
- A high-security subsystem addresses message protection, authentication, delegation, and authorization. Comprising both a set of service implementations (server programs at the bottom of Figure 4.5) and associated client libraries at the top.
- GT4 provides both web services and non-WS applications. The horizontal boxes in the client domain denote custom applications and/or third-party tools that access GT4 services.
- The toolkit programs provide a set of useful infrastructure services.

Three containers are used to host user-developed services written in Java, Python, and C, respectively. These containers provide implementations of security, management, discovery, state management, and other mechanisms frequently required when building services. They extend open source service hosting environments with support for a range of useful web service specifications, including WSRF, WS-Notification, and WS-Security.

A set of client libraries allow client programs in Java, C, and Python to invoke operations on both GT4 and user-developed services. The use of uniform abstractions and mechanisms means clients can interact with different services in similar ways, which facilitates construction of complex, interoperable systems and encourages code reuse.

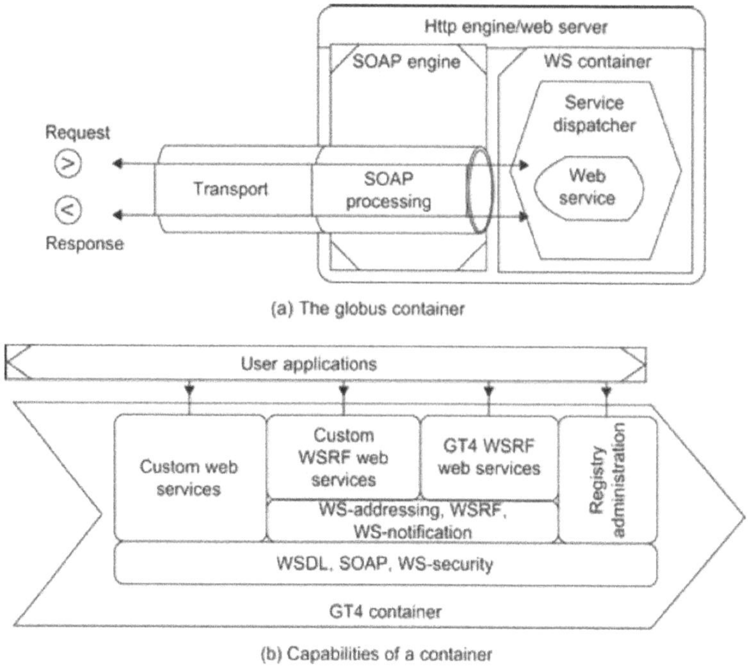

Figure 4.5 Globus container serving as a runtime environment for implementing web services in a grid platform.

4.4.3 Containers and Resources/Data Management

GRAM supports dynamic job execution with coordinated file staging. MDS is used for monitoring and discovery of available grid resources in a grid execution environment.

4.4.3.1 Resource Management Using GRAM

The GRAM module supports

- Web services for initiating, monitoring, and managing execution of computational jobs on remote computers in an open grid.
  GRAM is built atop various local resource allocation services.
- A standardized GRAM interface enables access to a variety of local resource management tools, such as the Load Sharing Facility the Network Queuing Environment, IBM's LoadLeveler, and Condor.
- This interface allows a client to express the resource type and quantity, data to be staged to and from the execution site, executable code and its arguments, security credentials to be used, and job persistence requirements.

Other operations enable clients to monitor the status of both allocated resources and running tasks and notify users about their status, as well as guide job execution on the grid.

The heart of GRAM contains a set of web services designed to run the Globus Web Services Resource Framework (WSRF) core hosting environment.

- Each submitted job is exposed with a distinct resource qualifying the requested service.
- The service provides an interface to monitor the status of the job or to terminate the job.
- Each compute element is accessed through a local scheduler.
- The service provides an interface to create the managed job resources to perform a job in that local scheduler.

4.4.3.2 Globus Container: A Runtime Environment

The Globus Container provides a basic runtime environment for hosting the web services needed to execute grid jobs.

- Figure 4.5(a) shows the container concept. The container is built with heavy use of SOAP engines.
- The main SOAP functions performed include the transport of incoming job requests and corresponding responses.
- It is also common to partition the hosting environment logic for transporting the SOAP message via an HTTP engine or web server.
  Figure 4.5(b) summarizes the capabilities of the containers.
- All WS containers implement the SOAP commands over the HTTP engine as a message transport protocol.
- Both transport-level and message-level security is enforced in all communications. The WS-addressing, WSRF, and WSNotification functions are implemented.
- The container supports logging using the Jakarta logging API.
- The container defines WSRF WS-Resources for grid services deployed in the container. Thus, any GT4 container can host many services to multiple jobs simultaneously.
- The container supports custom web services and host services whose client interfaces make use of WSRF and related mechanisms.
- The container also hosts advanced services provided by GT4, such as GRAM, MDS, and RFT.
- The application clients use the Globus container registry interfaces to determine which services are hosted in a particular container. The container administration interfaces are used to perform routine management functions.

4.4.3.3 Data Management Using GT4

Grid applications often need to provide access to and/or integrate large quantities of data at multiple sites. The GT4 tools can be used individually or in conjunction with other tools to develop interesting solutions to efficient data access. The following list briefly introduces these GT4 tools:

1. **GridFTP**
   - It supports reliable, secure, and fast memory-to-memory and disk-to-disk data movement over high-bandwidth WANs.
   - GridFTP adds additional features such as parallel data transfer, third-party data transfer, and striped data transfer

- GridFTP benefits from using the strong Globus Security Infrastructure for securing data channels with authentication and reusability.
- It has been reported that the grid has achieved 27 Gbit/second end-to-end transfer speeds over some WANs.

2. **RFT** provides reliable management of multiple GridFTP transfers. It has been used to orchestrate the transfer of millions of files among many sites simultaneously.
3. **RLS (Replica Location Service)** is a scalable system for maintaining and providing access to information about the location of replicated files and data sets.
4. **OGSA-DAI (Globus Data Access and Integration)** tools were developed by the UK e-Science program and provide access to relational and XML databases.

4.4.3.4 The MDS Services

Monitoring and discovery are two vital functions in any distributed system.

- Both tasks require the ability to collect information from multiple distributed information sources.
- GT4 provides monitoring and discovery support at a fundamental level.
- Figure 4.6 shows the Globus MDS infrastructure.
- By implementing WSRF and WS-Notification specifications, GT4 enables the association of XML-based resource properties with network entities.
- Multiple containers are built to carry out MDS indexing, GridFTP, and other functions needed to conduct monitoring and discovery of available resources.
- Grid services can be registered with distributed containers.

GT4 provides two aggregator services that collect data from any information source, whether XML-based or otherwise. In addition, GT4 provides a range of browser-based interfaces, command-line tools, and web service interfaces that allow users to query and access the collected information. In particular, the WebMDS service is configured via XSLT transformations to create specialized views of the index data. These mechanisms provide a powerful framework for monitoring diverse collections of distributed components.

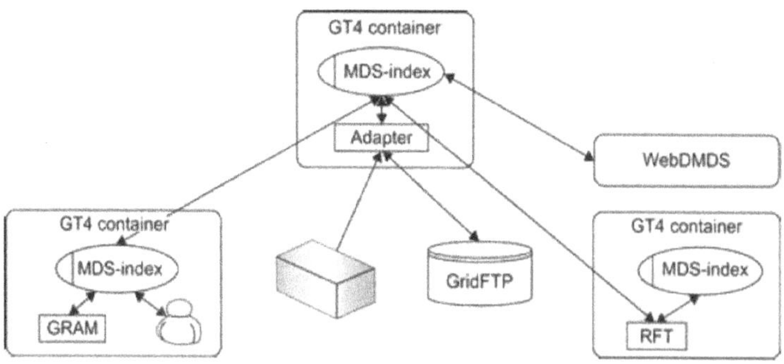

Figure 4.6 GT4 system monitoring and resource discovery infrastructure.

4.5 Introducing Hadoop

Hadoop is the Apache Software Foundation top-level project that holds the various Hadoop subprojects that graduated from the Apache Incubator. The Hadoop project provides and supports the development of open source software that supplies a framework for the development of highly scalable distributed computing applications. The Hadoop framework handles the processing details, leaving developers free to focus on application logic.

The introduction on the Hadoop project states:

The Apache Hadoop project develops open-source software for reliable, scalable, distributed computing, including:

Hadoop Core, our flagship sub-project, provides a distributed filesystem (HDFS) and support for the MapReduce distributed computing metaphor.

HBase builds on Hadoop Core to provide a scalable, distributed database.

Pig is a high-level data-flow language and execution framework for parallel computation.

It is built on top of Hadoop Core.

ZooKeeper is a highly available and reliable coordination system. Distributed applications use ZooKeeper to store and mediate updates for critical shared state.

Hive is a data warehouse infrastructure built on Hadoop Core that provides data summarization, adhoc querying and analysis of datasets.

The Hadoop Core project provides the basic services for building a cloud computing environment with commodity hardware, and the APIs for developing software that will run on that cloud. The two fundamental pieces of Hadoop Core are the MapReduce framework,

- the cloud computing environment,
- And he Hadoop Distributed File System (HDFS).

The Hadoop Core MapReduce framework requires a shared file system.

- This shared file system does not need to be a system-level file system, as long as there is a distributed file system plug-in available to the framework.

While Hadoop Core provides HDFS, HDFS is not required.

- In addition to HDFS, Hadoop Core supports the Cloud- Store file system and Amazon Simple Storage Service (S3) file system.
- The Hadoop Core framework comes with plug-ins for HDFS, CloudStore, and S3.
- Users are also free to use any distributed file system that is visible as a system-mounted file system, such as Network File System (NFS), Global File System (GFS), or Lustre.

When HDFS is used as the shared file system, Hadoop is able to take advantage of knowledge about which node hosts a physical copy of input data, and will attempt to schedule the task that is to read that data, to run on that machine.

4.5.1 Hadoop Core MapReduce

The Hadoop Distributed File System (HDFS)MapReduce environment provides the user with a sophisticated framework to manage the execution of map and reduce tasks across a cluster of machines. The user is required to tell the framework the following:

- The location(s) in the distributed file system of the job input
- The location(s) in the distributed file system for the job output
- The input format
- The output format
- The class containing the map function

- Optionally. the class containing the reduce function
- The JAR file(s) containing the map and reduce functions and any support classes

If a job does not need a reduce function, the user does not need to specify a reducer class, and a reduce phase of the job will not be run. The framework will partition the input, and schedule and execute map tasks across the cluster. If requested, it will sort the results of the map task and execute the reduce task(s) with the map output. The final output will be moved to the output directory, and the job status will be reported to the user.

MapReduce is oriented around key/value pairs.

- The framework will convert each record of **input** into a key/value pair, and each pair will be input to the map function once.
- The map **output** is a set of key/value pairs nominally one pair that is the transformed input pair, but it is perfectly acceptable to output multiple pairs. The map output pairs are grouped and sorted by key.
- The **reduce function** is called one time for each key, in sort sequence, with the key and the set of values that share that key.
- The reduce method may output an arbitrary number of key/value pairs, which are written to the output files in the job output directory.
- If the **reduce output keys** are unchanged from the reduce input keys, the final output will be sorted.

The framework provides two processes that handle the management of MapReduce jobs:

- TaskTracker manages the execution of individual map and reduce tasks on a compute node in the cluster.
- JobTracker accepts job submissions, provides job monitoring and control, and manages the distribution of tasks to the TaskTracker nodes.

Generally, there is one JobTracker process per cluster and one or more TaskTracker processes per node in the cluster. The JobTracker is a single point of failure, and the JobTracker will work around the failure of individual TaskTracker processes.

4.5.2 The Hadoop Distributed File System

HDFS is a file system that is designed for use for MapReduce jobs that read input in large chunks of input, process it, and write potentially large chunks of output. For reliability, file data is simply mirrored to multiple storage nodes. This is referred to as **replication in the Hadoop community**. As long as at least one replica of a data chunk is available, the consumer of that data will not know of storage server failures.

HDFS services are provided by two processes:

- NameNode handles management of the file system metadata, and provides management
  and control services.
- DataNode provides block storage and retrieval services.

There will be one NameNode process in an HDFS file system, and this is a single point of failure. Hadoop Core provides **recovery and automatic backup** of the NameNode, but no hot failover services. There will be multiple DataNode processes within the cluster, with typically one DataNode process per storage node in a cluster.

4.6 Introduction to Hadoop MapReduce

Hadoop supports the MapReduce model, which was introduced by Google as a method of solving a class of petascale problems with large clusters of inexpensive machines. The model is based on two distinct steps for an application:

- **Map:** An initial ingestion and transformation step, in which individual input records can be processed in parallel.
- **Reduce:** an aggregation or summarization step, in which all associated records, must be processed together by a single entity.

The core concept of MapReduce in Hadoop is that input may be split into logical chunks, and each chunk may be initially processed independently, by a map task. The results of these individual processing chunks can be physically partitioned into distinct sets, which are then sorted. Each sorted chunk is passed to a reduce task. Figure 4.7 illustrates how the MapReduce model works.

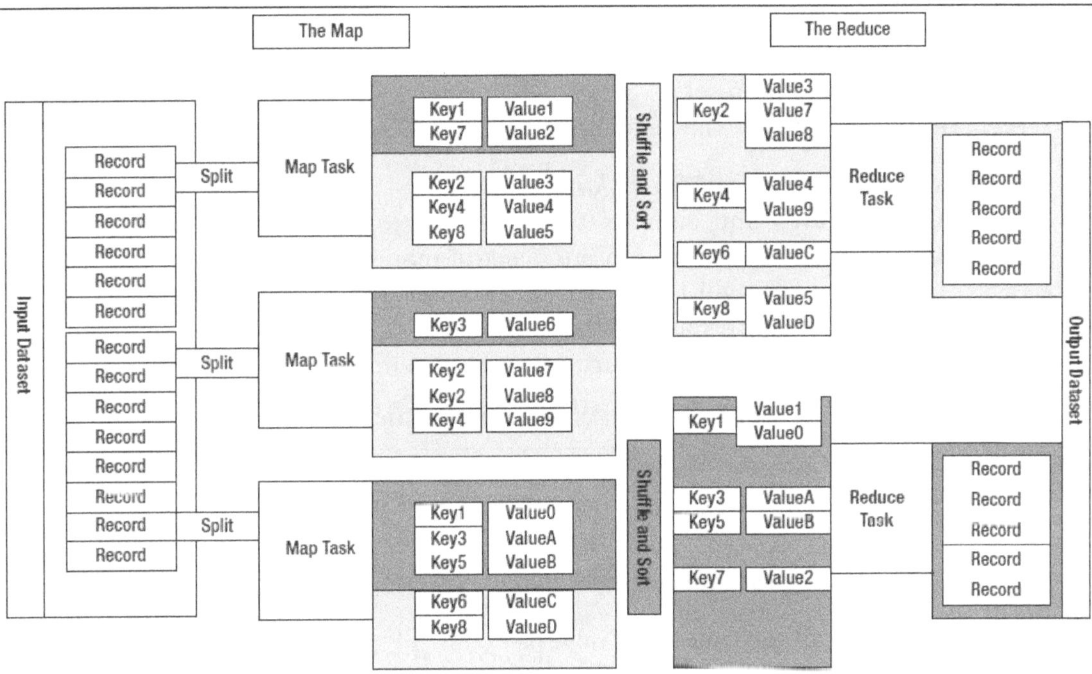

Figure 4.7 The MapReduce model

A **map task** may run on any compute node in the cluster, and multiple map tasks may be running in parallel across the cluster.

- The map task is responsible for transforming the input records into key/value pairs.
- The output of all of the maps will be partitioned, and each partition will be sorted.
- There will be one partition for each **reduce task**. Each partition's sorted keys and the values associated with the keys are then processed by the reduce task. There may be multiple reduce tasks running in parallel on the cluster.

The application developer needs to provide only four items to the Hadoop framework: the class that will read the input records and transform them into one key/value pair per record, a map method, a reduce method, and a class that will transform the key/value pairs that the reduce method outputs into output records.

My first MapReduce application was a specialized web crawler. This crawler received as input large sets of media URLs that were to have their content fetched and processed. The media items were large, and fetching them had a significant cost in time and resources.

The job had several steps:

1. Ingest the URLs and their associated metadata.
2. Normalize the URLs.
3. Eliminate duplicate URLs.
4. Filter the URLs against a set of exclusion and inclusion filters.
5. Filter the URLs against a do not fetch list.
6. Filter the URLs against a recently seen set.
7. Fetch the URLs.
8. Fingerprint the content items.
9. Update the recently seen set.
10. Prepare the work list for the next application.

4.6.1 The Parts of a Hadoop MapReduce Job

The user configures and submits a MapReduce job (or just job for short) to the framework, which will decompose the job into a set of map tasks, shuffles, a sort, and a set of reduce tasks. The framework will then manage the distribution and execution of the tasks, collect the output, and report the status to the user.

The job consists of the parts shown in Figure 4.8 and listed in Table 4.1.

Table 4.1. Parts of a MapReduce Job

Part	**Handled By**
Configuration of the job	User
Input splitting and distribution	Hadoop framework
Start of the individual map tasks with their input split	Hadoop framework
Map function, called once for each input key/value pair	User
Shuffle, which partitions and sorts the per-map output	Hadoop framework
Sort, which merge sorts the shuffle output for each partition of all map outputs	Hadoop framework
Start of the individual reduce tasks, with their input partition	Hadoop framework
Reduce function, which is called once for each unique input key, with all of the input values that share that key	User
Collection of the output and storage in the configured job output directory, in N parts, where N is the number of reduce tasks	Hadoop framework

The user is responsible for handling the job setup, specifying the input location(s), specifying the input, and ensuring the input is in the expected format and location. The framework is responsible for distributing the job among the TaskTracker nodes of the cluster;

running the map, shuffle, sort, and reduce phases; placing the output in the output directory; and informing the user of the job-completion status.

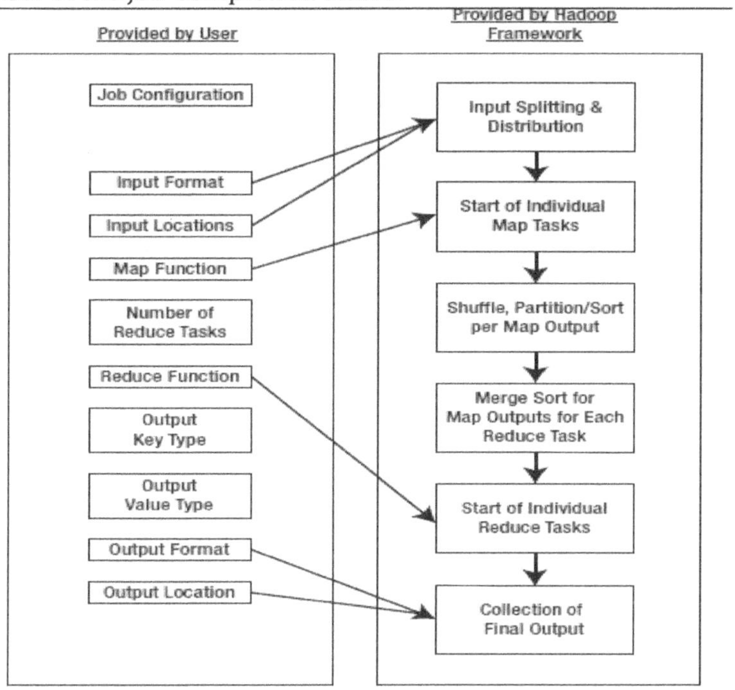

Figure 4.8 Parts of a MapReduce job

All the examples in this chapter are based on the file MapReduceIntro.java, shown in code 4.1. The job created by the code in MapReduceIntro.java will read all of its textual input line by line, and sort the lines based on that portion of the line before the first tab character. If there are no tab characters in the line, the sort will be based on the entire line. The MapReduceIntro.java file is structured to provide a simple example of configuring and running a MapReduce job.

Code 4.1. MapReduceIntro.java

```
package com.apress.hadoopbook.examples.ch2;
import java.io.IOException;
import org.apache.hadoop.io.Text;
import org.apache.hadoop.mapred.FileInputFormat;
import org.apache.hadoop.mapred.FileOutputFormat;
import org.apache.hadoop.mapred.JobClient;
import org.apache.hadoop.mapred.JobConf;
import org.apache.hadoop.mapred.KeyValueTextInputFormat;
import org.apache.hadoop.mapred.RunningJob;
```

```
import org.apache.hadoop.mapred.lib.IdentityMapper;
import org.apache.hadoop.mapred.lib.IdentityReducer;
import org.apache.log4j.Logger;
/** A very simple MapReduce example that reads textual input where
* each record is a single line, and sorts all of the input lines into
* a single output file.
*
* The records are parsed into Key and Value using the first TAB
* character as a separator. If there is no TAB character the entire
* line is the Key. *
*
* @author Jason Venner
*
*/
public class MapReduceIntro {
protected static Logger logger = Logger.getLogger(MapReduceIntro.class);
/**
* Configure and run the MapReduceIntro job.
*
* @param args
* Not used.
*/
public static void main(final String[] args) {
try {
/** Construct the job conf object that will be used to submit this job
* to the Hadoop framework. ensure that the jar or directory that
* contains MapReduceIntroConfig.class is made available to all of the
* Tasktracker nodes that will run maps or reduces for this job.
*/
final JobConf conf = new JobConf(MapReduceIntro.class);
/**
* Take care of some housekeeping to ensure that this simple example
* job will run
*/
MapReduceIntroConfig.
exampleHouseKeeping(conf,
MapReduceIntroConfig.getInputDirectory(),
MapReduceIntroConfig.getOutputDirectory());
/**
* This section is the actual job configuration portion /**
* Configure the inputDirectory and the type of input. In this case
* we are stating that the input is text, and each record is a
* single line, and the first TAB is the separator between the key
* and the value of the record.
```

```
*/
conf.setInputFormat(KeyValueTextInputFormat.class);
FileInputFormat.setInputPaths(conf,
MapReduceIntroConfig.getInputDirectory());
/** Inform the framework that the mapper class will be the
* {@link IdentityMapper}. This class simply passes the
* input Key Value pairs directly to its output, which in
* our case will be the shuffle.
*/
conf.setMapperClass(IdentityMapper.class);
/** Configure the output of the job to go to the output
* directory. Inform the framework that the Output Key
* and Value classes will be {@link Text} and the output
* file format will {@link TextOutputFormat}. The
* TextOutput format class joins produces a record of
* output for each Key,Value pair, with the following
* format. Formatter.format( "%s\t%s%n", key.toString(),
* value.toString() );.
*
* In addition indicate to the framework that there will be
* 1 reduce. This results in all input keys being placed
* into the same, single, partition, and the final output
* being a single sorted file.
*/
FileOutputFormat.setOutputPath(conf,
MapReduceIntroConfig.getOutputDirectory());
conf.setOutputKeyClass(Text.class);
conf.setOutputValueClass(Text.class);
conf.setNumReduceTasks(1); /** Inform the framework that the reducer class will be the
{@link
* IdentityReducer}. This class simply writes an output record key,
* value record for each value in the key, valueset it receives as
* input. The value ordering is arbitrary.
*/
conf.setReducerClass(IdentityReducer.class);
logger .info("Launching the job.");
/** Send the job configuration to the framework and request that the
* job be run.
*/ final RunningJob job = JobClient.runJob(conf);
logger.info("The job has completed.");
if (!job.isSuccessful()) {
logger.error("The job failed.");
System.exit(1);
} logger.info("The job completed successfully.");
```

```
System.exit(0);
} catch (final IOException e) { logger.error("The job has failed due to an IO Exception", e);
e.printStackTrace();}}}
```

4.6.1.1 Input Splitting

For the framework to be able to distribute pieces of the job to multiple machines, it needs to fragment the input into individual pieces, which can in turn be provided as input to the individual distributed tasks. Each fragment of input is called an **input split**. The default rules for how input splits are constructed from the actual input files are a combination of configuration parameters and the capabilities of the class that actually reads the input records.

An input split will normally be a contiguous group of records from a single input file, and in this case, there will be at least N input splits, where N is the number of input files. If the number of requested map tasks is larger than this number, or the individual files are larger than the suggested fragment size, there may be multiple input splits constructed of each input file. The user has considerable control over the number of input splits. The number and size of the input splits strongly influence overall job performance.

4.6.1.2 A Simple Map Function: IdentityMapper

The Hadoop framework provides a very simple map function, called IdentityMapper. It is used in jobs that only need to reduce the input, and not transform the raw input. We are going to examine the code of the IdentityMapper class, shown in code 4.2, in this section

Code 4.2. IdentityMapper.java

```
/**
* Licensed to the Apache Software Foundation (ASF) under one
* or more contributor license agreements. See the NOTICE file
* distributed with this work for additional information
* regarding copyright ownership. The ASF licenses this file
* to you under the Apache License, Version 2.0 (the
* "License"); you may not use this file except in compliance
* with the License. You may obtain a copy of the License at
*
* http://www.apache.org/licenses/LICENSE-2.0
*
* Unless required by applicable law or agreed to in writing, software
* distributed under the License is distributed on an "AS IS" BASIS,
* WITHOUT WARRANTIES OR CONDITIONS OF ANY KIND, either express or implied.
* See the License for the specific language governing permissions and
* limitations under the License.
*/
package org.apache.hadoop.mapred.lib;
import java.io.IOException;
import org.apache.hadoop.mapred.Mapper;
```

```
import org.apache.hadoop.mapred.OutputCollector;
import org.apache.hadoop.mapred.Reporter;
import org.apache.hadoop.mapred.MapReduceBase;
/** Implements the identity function, mapping inputs directly to outputs. */
public class IdentityMapper<K, V>
extends MapReduceBase implements Mapper<K, V, K, V> {
/** The identify function. Input key/value pair is written directly to
* output.*/
public void map(K key, V val,
OutputCollector<K, V> output, Reporter reporter)
throws IOException {
output.collect(key, val);
}
}
```

The magic piece of code is the line output.collect(key, val), which passes a key/value pair back to the framework for further processing. All map functions must implement the Mapper interface, which guarantees that the map function will always be called with a key. The key is an instance of a WritableComparable object, a value that is an instance of a Writable object, an output object, and a reporter.

4.6.1.3 A Simple Reduce Function: IdentityReducer

The Hadoop framework calls the reduce function one time for each unique key. The framework provides the key and the set of values that share that key. The framework-supplied class IdentityReducer is a simple example that produces one output record for every value. Code 4.3 shows this class.

Code4.3. IdentityReducer.java

```
/**
* Licensed to the Apache Software Foundation (ASF) under one
* or more contributor license agreements. See the NOTICE file
* distributed with this work for additional information
* regarding copyright ownership. The ASF licenses this file
* to you under the Apache License, Version 2.0 (the
* "License"); you may not use this file except in compliance
* with the License. You may obtain a copy of the License at
*
* http://www.apache.org/licenses/LICENSE-2.0
*
* Unless required by applicable law or agreed to in writing, software
* distributed under the License is distributed on an "AS IS" BASIS,
* WITHOUT WARRANTIES OR CONDITIONS OF ANY KIND, either express or implied.
```

```
* See the License for the specific language governing permissions and
* limitations under the License.
*/
package org.apache.hadoop.mapred.lib;
import java.io.IOException;
import java.util.Iterator;
import org.apache.hadoop.mapred.Reducer;
import org.apache.hadoop.mapred.OutputCollector;
import org.apache.hadoop.mapred.Reporter;
import org.apache.hadoop.mapred.MapReduceBase;
/** Performs no reduction, writing all input values directly to the output.
public class IdentityReducer<K, V>
extends MapReduceBase implements Reducer<K, V, K, V> {
/** Writes all keys and values directly to output. */
public void reduce(K key, Iterator<V> values,
OutputCollector<K, V> output, Reporter reporter)
throws IOException {
while (values.hasNext()) {
output.collect(key, values.next());
}
}
```

If you require the output of your job to be sorted, the reducer function must pass the key objects to the output.collect() method unchanged. The reduce phase is, however, free to output any number of records, including zero records, with the same key and different values. This particular constraint is also why the map tasks may be multithreaded, while the reduce tasks are explicitly only single-threaded.

4.6.2 Configuring a Job

All Hadoop jobs have a driver program that configures the actual MapReduce job and submits it to the Hadoop framework. This configuration is handled through the JobConf object. The sample class MapReduceIntro provides a walk-through for using the JobConf object to configure and submit a job to the Hadoop framework for execution. The code relies on a class called MapReduceIntroConfig, shown in code 4.4, which ensures that the input and output directories are set up and ready.

Code 4.4. MapReduceIntroConfig.java

```
package com.apress.hadoopbook.examples.ch2;
import java.io.IOException;
import java.util.Formatter;
```

```
import java.util.Random;
import org.apache.hadoop.fs.FSDataOutputStream;
import org.apache.hadoop.fs.FileStatus;
import org.apache.hadoop.fs.FileSystem;
import org.apache.hadoop.fs.Path;
import org.apache.hadoop.mapred.JobConf;
import org.apache.log4j.Logger;
/** A simple class to handle the housekeeping for the MapReduceIntro
* example job.
*
*
* <p>
* This job explicitly configures the job to run, locally and without a
* distributed file system, as a stand alone application.
* </p>
* <p>
* The input is read from the directory /tmp/MapReduceIntroInput and
* the output is written to the directory
* /tmp/MapReduceIntroOutput. If the directory
* /tmp/MapReduceIntroInput is missing or empty, it is created and
* some input data files generated. If the directory
* /tmp/MapReduceIntroOutput is present, it is removed.
* </p>
*
* @author Jason Venner
*/
public class MapReduceIntroConfig {
/**
* Log4j is the recommended way to provide textual information to the user
* about the job.
*/
protected static Logger logger =
Logger.getLogger(MapReduceIntroConfig.class);
/** Some simple defaults for the job input and job output. */
/**
* This is the directory that the framework will look for input files in.
* The search is recursive if the entry is a directory.
*/
protected static Path inputDirectory =
new Path("file:///tmp/MapReduceIntroInput");
/**
* This is the directory that the job output will be written to. It must not
* exist at Job Submission time.
*/
```

```
protected static Path outputDirectory =
new Path("file:///tmp/MapReduceIntroOutput");
/**
* Ensure that there is some input in the <code>inputDirectory</code>,
* the <code>outputDirectory</code> does not exist and that this job will
* be run as a local stand alone application.
*
* @param conf
* The {@link JobConf} object that is required for doing file
* system access.
* @param inputDirectory
* The directory the input will reside in.
* @param outputDirectory
* The directory that the output will reside in
* @throws IOException
*/
protected static void exampleHouseKeeping(final JobConf conf,
final Path inputDirectory, final Path outputDirectory)
throws IOException {
/** * Ensure that this job will be run stand alone rather than relying on
* the services of an external JobTracker.
*/ conf.set("mapred.job.tracker", "local");
/** Ensure that no global file system is required to run this job. */
conf.set("fs.default.name", "file:///");
/**
* Reduce the in ram sort space, so that the user does not need to
* increase the jvm memory size. This sets the sort space to 1 Mbyte,
* which is very small for a real job.
*/
conf.setInt("io.sort.mb", 1);
/**
* Generate some sample input if the <code>inputDirectory</code> is
* empty or absent.
*/
generateSampleInputIf(conf, inputDirectory);
/**
* Remove the file system item at <code>outputDirectory</code> if it
* exists.
*/
if (!removeIf(conf, outputDirectory)) {
logger.error("Unable to remove " + outputDirectory + "job aborted");
System.exit(1);
}
}
```

```
/**
* Generate <code>fileCount</code> files in the directory
* <code>inputDirectory</code>, where the individual lines of the file
* are a random integer TAB file name.
*
* The file names will be file-N where N is between 0 and
* <code>fileCount</code> - 1. There will be between 1 and
* <code>maxLines</code> + 1 lines in each file.
*
* @param fs
* The file system that <code>inputDirectory</code> exists in.
* @param inputDirectory
* The directory to create the files in. This directory must
* already exist.
* @param fileCount
* The number of files to create.
* @param maxLines
* The maximum number of lines to write to the file.
*protected static void generateRandomFiles(final FileSystem fs,
final Path inputDirectory, final int fileCount, final int maxLines)
throws IOException {
final Random random = new Random();
logger .info( "Generating 3 input files of random data," +
"each record is a random number TAB the input file name");
for (int file = 0; file < fileCount; file++) {
final Path outputFile = new Path(inputDirectory, "file-" + file);
final String qualifiedOutputFile = outputFile.makeQualified(fs)
.toUri().toASCIIString();
FSDataOutputStream out = null;
try {
/**
* This is the standard way to create a file using the Hadoop
* Framework. An error will be thrown if the file already
* exists.
*/
out = fs.create(outputFile);
final Formatter fmt = new Formatter(out);
final int lineCount = (int) (Math.abs(random.nextFloat())
* maxLines + 1);
for (int line = 0; line < lineCount; line++) {
fmt.format("%d\t%s%n", Math.abs(random.nextInt()),
qualifiedOutputFile);
}
fmt.flush();
```

```
} finally {
/**
* It is very important to ensure that file descriptors are
* closed. The distributed file system code can run out of file
* descriptors and the errors generated in that case are
* misleading.
*/
out.close();
}
}
}
/**
* This method will generate some sample input, if the
* <code>inputDirectory</code> is missing or empty.
*
* This method also demonstrates some of the basic APIs for interacting
* with file systems and files. Note: the code has no particular knowledge
* of the type of file system.
*
* @param conf
* The Job Configuration object, used for acquiring the
* {@link FileSystem} objects.
* @param inputDirectory
* The directory to ensure has sample files.
* @throws IOException
*/
protected static void generateSampleInputIf(final JobConf conf,
final Path inputDirectory) throws IOException {
boolean inputDirectoryExists;
final FileSystem fs = inputDirectory.getFileSystem(conf);
if ((inputDirectoryExists = fs.exists(inputDirectory))
&& !isEmptyDirectory(fs, inputDirectory)) {
if (logger.isDebugEnabled()) {
logger
.debug("The inputDirectory "
+ inputDirectory
+ " exists and is either a"
+ " file or a non empty directory");
}
return;
}
/**
* We should only get here if <code>inputDirectory</code> does not
* exist, or is an empty directory.
```

```
*/
if (!inputDirectoryExists) {
if (!fs.mkdirs(inputDirectory)) {
logger.error("Unable to make the inputDirectory "
+ inputDirectory.makeQualified(fs) + " aborting job");
System.exit(1);
}
}final int fileCount = 3;
final int maxLines = 100;
generateRandomFiles(fs, inputDirectory, fileCount, maxLines);
}
/**
* bean access getter to the {@link #inputDirectory} field.
*
* @return the value of inputDirectory.
*/
public static Path getInputDirectory() {
return inputDirectory;
}
/**
* bean access getter to the {@link outputDirectory} field.
*
* @return the value of outputDirectory.
*/
public static Path getOutputDirectory() {
return outputDirectory;
}
/**
* Determine if a directory has any non zero files in it or its descendant
* directories.
*
* @param fs
* The {@link FileSystem} object to use for access.
* @param inputDirectory
* The root of the directory tree to search
* @return true if the directory is missing or does not contain at least one
* non empty file.
* @throws IOException
*/
private static boolean isEmptyDirectory(final FileSystem fs,
final Path inputDirectory) throws IOException {
/**
* This is the standard way to read a directory's contents. This can be
* quite expensive for a large directory.
```

```
*/
final FileStatus[] statai = fs.listStatus(inputDirectory); /**
* This method returns null under some circumstances, in particular if
* the directory does not exist.
*/
if ((statai == null) || (statai.length == 0)) {
if (logger.isDebugEnabled()) {
logger.debug(inputDirectory.makeQualified(fs).toUri()
+ " is empty or missing");
}
return true;
}
if (logger.isDebugEnabled()) {
logger.debug(inputDirectory.makeQualified(fs).toUri()
+ " is not empty");
}
/** Try to find a file in the top level that is not empty. */
for (final FileStatus status : statai) {
if (!status.isDir() && (status.getLen() != 0)) {
if (logger.isDebugEnabled()) {
logger.debug("A non empty file "
+ status.getPath().makeQualified(fs).toUri()
+ " was found");
return false;
}
}
}
/** Recurse if there are sub directories,
* looking for a non empty file.
*/
for (final FileStatus status : statai) {
if (status.isDir() && isEmptyDirectory(fs, status.getPath())) {
continue;
}
/**
* If status is a directory it must not be empty or the previous
* test block would have triggered.
*/
if (status.isDir()) {
return false;
}
}
/**
* Only get here if no non empty files were found in the entire subtree
```

```
* of <code>inputPath</code>.
*/
return true; /**
* Ensure that the <code>outputDirectory</code> does not exist.
*
* <p>
* The framework requires that the output directory not be present at job
* submission time.
* </p>
* <p>
* This method also demonstrates how to remove a directory using the
* {@link FileSystem} API.
* </p>
*
* @param conf
* The configuration object. This is needed to know what file
* systems and file system plugins are being used.
* @param outputDirectory
* The directory that must be removed if present.
* @return true if the the <code>outputPath</code> is now missing, or
* false if the <code>outputPath</code> is present and was unable
* to be removed.
* @throws IOException
* If there is an error loading or configuring the FileSystem
* plugin, or other IO error when attempting to access or remove
* the <code>outputDirectory</code>.
*/
protected static boolean removeIf(final JobConf conf,
final Path outputDirectory) throws IOException {
/** This is standard way to acquire a FileSystem object. */
final FileSystem fs = outputDirectory.getFileSystem(conf);
/**
* If the <code>outputDirectory</code> does not exist this method is
* done.
*/
if (!fs.exists(outputDirectory)) {
if (logger.isDebugEnabled()) {
logger .debug("The output directory does not exist,"
+ " no removal needed.");
}
return true;
}
/**
* The getFileStatus command will throw an IOException if the path does
```

```
* notexist.
*/final FileStatus status = fs.getFileStatus(outputDirectory);
logger.info("The job output directory "
+ outputDirectory.makeQualified(fs) + " exists"
+ (status.isDir() ? " and is not a directory" : "")
+ " and will be removed");
/**
* Attempt to delete the file or directory. delete recursively just in
* case <code>outputDirectory</code> is a directory with
* sub-directories.
*/
if (!fs.delete(outputDirectory, true)) {
logger.error("Unable to delete the configured output directory "
+ outputDirectory);
return false;
}
/** The outputDirectory did exist, but has now been removed. */
return true;
}
/**
* bean access setter to the {@link inputDirectory} field.
*
* @param inputDirectory
* The value to set inputDirectory to.
*/
public static void setInputDirectory(final Path inputDirectory) {
MapReduceIntroConfig.inputDirectory = inputDirectory;
}
/**
* bean access setter for the {@link outpuDirectory field.
*
* @param outputDirectory
* The value to set outputDirectory to.
*/
public static void setOutputDirectory(final Path outputDirectory) {
MapReduceIntroConfig.outputDirectory = outputDirectory;
}
}
```

First, you must create a JobConf object. It is good practice to pass in a class that is contained in the JAR file that has your map and reduce functions. This ensures that the framework will make the JAR available to the map and reduce tasks run for your job.

```
JobConf conf = new JobConf(MapReduceIntro.class);
```

Now that you have a JobConfig object, conf, you need to set the required parameters for the job. These include the input and output directory locations, the format of the input and output, and the mapper and reducer classes. All jobs will have a map phase, and the map phase is responsible for handling the job input. The configuration of the map phase requires you to specify the input locations and the class that will produce the key/value pairs from the input, the mapper class, and potentially, the suggested number of map tasks, map output types, and per-map task threading, as listed in Table 4.2.

Most Hadoop Core jobs have their input as some set of files, and these files are either a textual key/value pair per line or a Hadoop-specific binary file format that provides serialized key/value pairs. The class that handles the key/value text input is KeyValueTextInputFormat. The class that handles the Hadoop-specific binary file is SequenceFileInputFormat.

Table 4.2 Map Phase Configuration

Element	Required?	Default
Input path(s)	Yes	
Class to read and convert the input path elements to key/value pairs	Yes	
Map output key class	No	Job output key class
Map output value class	No	Job output value class
Class supplying the map function	Yes	
Suggested minimum number of map tasks	No	Cluster default
Number of threads to run in each map task	No	1

4.6.2.1 Specifying Input Formats

The Hadoop framework provides a large variety of input formats. The major distinctions are between textual input formats and binary input formats. The following are the available formats:

- KeyValueTextInputFormat: Key/value pairs, one per line.
- TextInputFormant: The key is the line number, and the value is the line.
- NLineInputFormat: Similar to KeyValueTextInputFormat, but the splits are based on N lines of input rather than Y bytes of input.
- MultiFileInputFormat: An abstract class that lets the user implement an input format that aggregates multiple files into one split.
- SequenceFIleInputFormat: The input file is a Hadoop sequence file, containing serialized key/value pairs.

KeyValueTextInputFormat and SequenceFileInputFormat are the most commonly used input formats. The examples in this chapter use KeyValueTextInputFormat, as the input files are human-readable. The following block of code informs the framework of the type and location of the job input:

```
/**
* This section is the actual job configuration portion /**
* Configure the inputDirectory and the type of input. In this case
* we are stating that the input is text, and each record is a
* single line, and the first TAB is the separator between the key
* and the value of the record.
*/
conf.setInputFormat(KeyValueTextInputFormat.class);
FileInputFormat.setInputPaths(conf,
MapReduceIntroConfig.getInputDirectory());
```

The line conf.setInputFormat(KeyValueTextInputFormat.class) informs the framework that all of the files used for input will be textual key/value pairs, one per line.

The framework knows where to look for the input files and the class to use to generate key/value pairs from the input files, you need to inform the framework which map function to use.

```
/** Inform the framework that the mapper class will be the {@link
* IdentityMapper}. This class simply passes the input key-value
* pairs directly to its output, which in our case will be the
* shuffle.
*/
conf.setMapperClass(IdentityMapper.class);
```

4.6.2.2 Setting the Output Parameters

The framework requires that the output parameters be configured, even if the job will not produce any output. The framework will collect the output from the specified tasks (either the output of the map tasks for a MapReduce job that did not include reduce tasks or the output of the job's reduce tasks) and place them into the configured output directory. To avoid issues with file name collisions when placing the task output into the output directory, the framework requires that the output directory not exist when you start the job.

In our simple example, the MapReduceIntroConfig class handles ensuring that the output directory does not exist and provides the path to the output directory. The output parameters are actually a little more comprehensive than just the setting of the output path. The code will also set the output format and the output key and value classes.

The Text class is the functional equivalent of a String. It implements the WritableComparable interface, which is necessary for keys, and the Writable interface (which is actually a subset of WritableComparable), which is necessary for values. Unlike String, Text is mutable, and the Text class has some explicit methods for UTF-8 byte handling.

The key feature of a Writable is that the framework knows how to serialize and deserialize a Writable object. The WritableComparable adds the compareTo interface so the framework knows how to sort the WritableComparable objects. The interface references for Writable Comparable and Writable are shown in code 4.5 and 4.6.

The following code block provides an example of the minimum required configuration for the output of a MapReduce job:

```
/** Configure the output of the job to go to the output directory.
```

```
* Inform the framework that the Output Key and Value classes will be
* {@link Text} and the output file format will {@link
* TextOutputFormat}. The TextOutput format class produces a record of
* output for each Key,Value pair, with the following format.
* Formatter.format( "%s\t%s%n", key.toString(), value.toString() );.
*
* In addition indicate to the framework that there will be
* 1 reduce. This results in all input keys being placed
* into the same, single, partition, and the final output
* being a single sorted file.
*/
FileOutputFormat.setOutputPath(conf,
MapReduceIntroConfig.getOutputDirectory());
conf.setOutputKeyClass(Text.class);
conf.setOutputValueClass(Text.class);
```

The File Output Format. set Output Path(conf, MapReduceIntroConfig.getOutputDirectory()) setting is familiar from the input example discussed earlier in the chapter. The conf.setOutputKeyClass(Text.class) and conf.setOutputValueClass(Text.class) settings are new. These settings inform the framework of the types of the key/value pairs to expect for the reduce phase. The method to set the output key class for the map output is conf.setMapOutputKeyClass(Class<? extends WritableComparable>). To set the output value class, the method is conf.setMapOutputValueClass(Class<? extends Writable>).

Code 4.5. WritableComparable.java	Listing 2-6. Writable.java
/** * Licensed to the Apache Software Foundation (ASF) under one * or more contributor license agreements. See the NOTICE file * distributed with this work for additional information * regarding copyright ownership. The ASF licenses this file * to you under the Apache License, Version 2.0 (the * "License"); you may not use this file except in compliance * with the License. You may obtain a copy of the License at * * http://www.apache.org/licenses/LICENSE-2.0	/** * Licensed to the Apache Software Foundation (ASF) under one * or more contributor license agreements. See the NOTICE file * distributed with this work for additional information * regarding copyright ownership. The ASF licenses this file* to you under the Apache License, Version 2.0 (the * "License"); you may not use this file except in compliance * with the License. You may obtain a copy of the License at * * http://www.apache.org/licenses/LICENSE-2.0 *

```
*
* Unless required by applicable law or agreed to in writing, software
* distributed under the License is distributed on an "AS IS" BASIS,
* WITHOUT WARRANTIES OR CONDITIONS OF ANY KIND, either express or implied.
* See the License for the specific language governing permissions and
* limitations under the License.
*/package org.apache.hadoop.io;
/**
* A {@link Writable} which is also {@link Comparable}.
*
* <p><code>WritableComparable</code>s can be compared to each other, typically
* via <code>Comparator</code>s. Any type which is to be used as a
* <code>key</code> in the Hadoop Map-Reduce framework should implement this
* interface.</p>
*
* <p>Example:</p>
* <p><blockquote><pre>
* public class MyWritableComparable implements WritableComparable {
* // Some data
* private int counter;
* private long timestamp;
*
* public void write(DataOutput out) throws IOException {
* out.writeInt(counter);
* out.writeLong(timestamp);
* }
*
* public void readFields(DataInput in) throws IOException {
* counter = in.readInt();
* timestamp = in.readLong();
* }
*
* public int
```

```
* Unless required by applicable law or agreed to in writing, software
* distributed under the License is distributed on an "AS IS" BASIS,
* WITHOUT WARRANTIES OR CONDITIONS OF ANY KIND, either express or implied.
* See the License for the specific language governing permissions and
* limitations under the License.
*/
package org.apache.hadoop.io;
import java.io.DataOutput;
import java.io.DataInput;
import java.io.IOException;
/*** A serializable object which implements a simple, efficient, serialization
* protocol, based on {@link DataInput} and {@link DataOutput}.
* <p>Any <code>key</code> or <code>value</code> type in the Hadoop Map-Reduce* framework implements this interface.</p>
** <p>Implementations typically implement a static <code>read(DataInput)</code>
* method which constructs a new instance, calls {@link #readFields(DataInput)}
* and returns the instance.</p>
** <p>Example:</p>* <p><blockquote><pre>
* public class MyWritable implements Writable {
* // Some data * private int counter;
* private long timestamp; *
* public void write(DataOutput out) throws IOException { * out.writeInt(counter);
* out.writeLong(timestamp); * }
* public void readFields(DataInput in) throws IOException { * counter = in.readInt();
* timestamp = in.readLong(); * }public static MyWritable read(DataInput in) throws OException { * MyWritable w = new MyWritable(); * w.readFields(in);
* return w;
* }* }* </pre></blockquote></p>
*/public interface Writable {
```

```
compareTo(MyWritableComparable w) {
* int thisValue = this.value;
* int thatValue = ((IntWritable)o).value;
* return (thisValue &lt; thatValue ? -1 :
(thisValue==thatValue ? 0 : 1));
* }
* }
* </pre></blockquote></p>
*/
public interface WritableComparable<T>
extends Writable, Comparable<T> {
}
```

```
/*** Serialize the fields of this object to
<code>out</code>.
** @param out <code>DataOuput</code> to
serialize this object into.
* @throws IOException
*/void write(DataOutput out) throws
IOException;
/*** Deserialize the fields of this object from
<code>in</code>.** <p>For efficiency,
implementations should attempt to re-use
storage in the* existing object where
possible.</p>
** @param in <code>DataInput</code> to
deseriablize this object from.* @throws
IOException*/void readFields(DataInput in)
throws IOException;}
*
```

4.6.2.3 Configuring the Reduce Phase

To configure the reduce phase, the user must supply the framework with five pieces of information:

- The number of reduce tasks; if zero, no reduce phase is run
- The class supplying the reduce method
- The input key and value types for the reduce task; by default, the same as the reduce output
- The output key and value types for the reduce task
- The output file type for the reduce task output

The input and output key and value types, as well as the output file type, are the same as those covered in the previous "Setting the Output Parameters" section. Here, we will look at setting the number of reduce tasks and the reducer class.

The configured number of reduce tasks determines the number of output files for a job that will run the reduce phase. Tuning this value will have a significant impact on the overall performance of your job. The time spent sorting the keys for each output file is a function of the number of keys. In addition, the number of reduce tasks determines the maximum number of reduce tasks that can be run in parallel.

The framework generally has a default number of reduce tasks configured. This value is set by the mapred.reduce.tasks parameter, which defaults to 1. This will result in a single output file containing all of the output keys, in sorted order. There will be one reduce task, run on a single machine that processes every key.

The number of reduce tasks is commonly set in the configuration phase of a job.

```
conf.setNumReduceTasks(1);
```

In general, unless there is a significant need for a single output file, the number of reduce tasks is set to roughly the number of simultaneous execution slots in the cluster. The class DataJoinReduceOutput is provided as a sample for efficiently merging multiple reduce task outputs into a single sorted file.

The reducer class needs to be set only if the number of reduce tasks is not zero. It is very common to not need a reducer, since frequently you do not require sorted output or value grouping by key. The actual setting of the reducer class is straightforward:

```
/** Inform the framework that the reducer class will be the
* {@link IdentityReducer}. This class simply writes an output record
* key/value record for each value in the key/value set it receives as
* input. The value ordering is arbitrary.
*/
conf.setReducerClass(IdentityReducer.class);
```

4.6.3 Running a Job

The ultimate aim of all your MapReduce job configuration is to actually run that job. The MapReduceIntro.java example code 4.1 demonstrates a common and simple way to run a job:

```
logger .info("Launching the job.");
/** Send the job configuration to the framework
* and request that the job be run.
*/
final RunningJob job = JobClient.runJob(conf);
logger.info("The job has completed.");
```

The method runJob() submits the configuration information to the framework and waits for the framework to finish running the job. The response is provided in the job object. The RunningJob class provides a number of methods for examining the response. Perhaps the most useful is job.isSuccessful().

The response should be as follows:

```
ch2.MapReduceIntroConfig: Generating 3 input files of random data, each record
is a random number TAB the input file name
ch2.MapReduceIntro: Launching the job.
jvm.JvmMetrics: Initializing JVM Metrics with processName=JobTracker, sessionId=
mapred.JobClient: Use GenericOptionsParser for parsing the arguments.
Applications should implement Tool for the same.
mapred.FileInputFormat: Total input paths to process : 3
mapred.FileInputFormat: Total input paths to process : 3
mapred.FileInputFormat: Total input paths to process : 3
mapred.FileInputFormat: Total input paths to process : 3
mapred.JobClient: Running job: job_local_0001
mapred.MapTask: numReduceTasks: 1
mapred.MapTask: io.sort.mb = 1
mapred.MapTask: data buffer = 796928/996160
mapred.MapTask: record buffer = 2620/3276
mapred.MapTask: Starting flush of map output
```

```
mapred.MapTask: bufstart = 0; bufend = 664; bufvoid = 996160
mapred.MapTask: kvstart = 0; kvend = 14; length = 3276
mapred.MapTask: Index: (0, 694, 694)
mapred.MapTask: Finished spill 0
mapred.LocalJobRunner: file:/tmp/MapReduceIntroInput/file-2:0+664
mapred.TaskRunner: Task 'attempt_local_0001_m_000000_0' done.
mapred.TaskRunner: Saved output of task 'attempt_local_0001_m_000000_0' to
file:/tmp/MapReduceIntroOutput
mapred.MapTask: numReduceTasks: 1
mapred.MapTask: io.sort.mb = 1
mapred.MapTask: data buffer = 796928/996160
mapred.MapTask: record buffer = 2620/3276
mapred.MapTask: Starting flush of map output
mapred.MapTask: bufstart = 0; bufend = 3418; bufvoid = 996160
mapred.MapTask: kvstart = 0; kvend = 72; length = 3276
mapred.MapTask: Index: (0, 3564, 3564)
mapred.MapTask: Finished spill 0
mapred.LocalJobRunner: file:/tmp/MapReduceIntroInput/file-1:0+3418
mapred.TaskRunner: Task 'attempt_local_0001_m_000001_0' done.
mapred.TaskRunner: Saved output of task 'attempt_local_0001_m_000001_0' to
file:/tmp/MapReduceIntroOutput
mapred.MapTask: numReduceTasks: 1
mapred.MapTask: io.sort.mb = 1
mapred.MapTask: data buffer = 796928/996160
mapred.MapTask: Starting flush of map output
mapred.MapTask: bufstart = 0; bufend = 3986; bufvoid = 996160
mapred.MapTask: kvstart = 0; kvend = 84; length = 3276
mapred.MapTask: Index: (0, 4156, 4156)
mapred.MapTask: Finished spill 0
mapred.LocalJobRunner: file:/tmp/MapReduceIntroInput/file-0:0+3986
mapred.TaskRunner: Task 'attempt_local_0001_m_000002_0' done.
mapred.TaskRunner: Saved output of task 'attempt_local_0001_m_000002_0' to
file:/tmp/MapReduceIntroOutput
mapred.ReduceTask: Initiating final on-disk merge with 3 files
mapred.Merger: Merging 3 sorted segments
mapred.Merger: Down to the last merge-pass, with 3 segments left of total size:
8414 bytes
mapred.LocalJobRunner: reduce > reduce
mapred.TaskRunner: Task 'attempt_local_0001_r_000000_0' done.
mapred.TaskRunner: Saved output of task 'attempt_local_0001_r_000000_0' to
file:/tmp/MapReduceIntroOutput
mapred.JobClient: Job complete: job_local_0001
mapred.JobClient: Counters: 11
mapred.JobClient: File Systems
```

```
mapred.JobClient: Local bytes read=230060
mapred.JobClient: Local bytes written=319797
mapred.JobClient: Map-Reduce Framework
mapred.JobClient: Reduce input groups=170
mapred.JobClient: Combine output records=0
mapred.JobClient: Map input records=170
mapred.JobClient: Reduce output records=170
mapred.JobClient: Map output bytes=8068
mapred.JobClient: Map input bytes=8068
mapred.JobClient: Combine input records=0
mapred.JobClient: Map output records=170
mapred.JobClient: Reduce input records=170
ch2.MapReduceIntro: The job has completed.
ch2.MapReduceIntro: The job completed successfully.
```

Congratulations, you have run a MapReduce job.

The single output file of the reduce task in the file /tmp/MapReduceIntroOutput/part-00000 will have a series of lines of the form Number TAB file:/tmp/MapReduceIntroInput/file-N. The first thing you will notice is that the numbers don't seem to be in order. The code that generates the input produces a random number for the key of each line, but the example tells the framework that the keys are Text. Therefore, the numbers have been sorted as text rather than as numbers.

4.7 The Hadoop Distributed File system (HDFS)

Hadoop comes with a distributed filesystem called HDFS, which stands for Hadoop Distributed Filesystem. HDFS is Hadoop's flagship file system but Hadoop actually has a general purpose file system abstraction, so we'll see along the way how Hadoop integrates with other storage systems (such as the local filesystem and Amazon S3).

4.7.1 The Design of HDFS

HDFS is a file system designed for storing very large files with streaming data access patterns, running on clusters of commodity hardware. Let's examine this statement in more detail:

- **Very large files-**"Very large" in this context means files that are hundreds of megabytes, gigabytes, or terabytes in size. There are Hadoop clusters running today that store petabytes of data.
- **Streaming data access-** HDFS is built around the idea that the most efficient data processing pattern is a write-once, read-many-times pattern. A dataset is typically generated or copied from source, and then various analyses are performed on that dataset over time. Each analysis will involve a large proportion, if not all, of the dataset, so the time to read the whole dataset is more important than the latency in reading the first record.
- **Commodity hardware-** Hadoop doesn't require expensive, highly reliable hardware. It's designed to run on clusters of commodity hardware (commonly available

hardware that can be obtained from multiple vendors)3 for which the chance of node failure across the cluster is high, at least for large clusters. HDFS is designed to carry on working without a noticeable interruption to the user in the face of such failure.

It is also worth examining the applications for which using HDFS does not work so well. Although this may change in the future, these are areas where HDFS is not a good fit today:

- **Low-latency data access-** Applications that require low-latency access to data, in the tens of milliseconds range, will not work well with HDFS. Remember, HDFS is optimized for delivering a high throughput of data, and this may be at the expense of latency. HBase is currently a better choice for low-latency access.
- **Lots of small files-** Because the name node holds file system metadata in memory, the limit to the number of files in a file system is governed by the amount of memory on the name node. As a rule of thumb, each file, directory, and block takes about 150 bytes. So, for example, if you had one million files, each taking one block, you would need at least 300 MB of memory. Although storing millions of files is feasible, billions is beyond the capability of current hardware.
- **Multiple writers, arbitrary file modifications-** Files in HDFS may be written to by a single writer. Writes are always made at the end of the file, in append-only fashion. There is no support for multiple writers or for modifications at arbitrary offsets in the file.

4.7.2 HDFS Concepts

Blocks

A disk has a block size, which is the minimum amount of data that it can read or write. HDFS has the concept of a block, but it is a much larger unit 128 MB by default. Like in a filesystem for a single disk, files in HDFS are broken into block-sized chunks, which are stored as independent units. Unlike a filesystem for a single disk, a file in HDFS that is smaller than a single block does not occupy a full block's worth of underlying storage. (For example, a 1 MB file stored with a block size of 128 MB uses 1 MB of disk space, not 128 MB.) When unqualified, the term "block" in this book refers to a block in HDFS.

- **Large blocks in HDFS**

HDFS blocks are large compared to disk blocks, and the reason is to minimize the cost of seeks. If the block is large enough, the time it takes to transfer the data from the disk can be significantly longer than the time to seek to the start of the block. Thus, transferring a large file made of multiple blocks operates at the disk transfer rate.

Having a block abstraction for a distributed file system brings several benefits.

- **A file can be larger than any single disk in the network**. There's nothing that requires the blocks from a file to be stored on the same disk, so they can take advantage of any of the disks in the cluster. In fact, it would be possible, if unusual, to store a single file on an HDFS cluster whose blocks filled all the disks in the cluster.
- **Making the unit of abstraction a block rather than a file simplifies the storage subsystem**. Simplicity is something to strive for in all systems, but it is especially important for a distributed system in which the failure modes are so varied. The storage subsystem deals with blocks, simplifying storage management.

- **Blocks fit well with replication for providing fault tolerance and availability**. To insure against corrupted blocks and disk and machine failure, each block is replicated to a small number of physically separate machines (typically three). If a block becomes unavailable, a copy can be read from another location in a way that is transparent to the client.

Namenodes and Datanodes

An HDFS cluster has two types of nodes operating in a master–worker pattern:

➔ **A namenode (the master)**:
  - The **namenode** manages the filesystem namespace.
  - It maintains the filesystem tree and the metadata for all the files and directories in the tree.
  - This information is stored persistently on the local disk in the form of two files: the namespace image and the edit log.

➔ **A number of datanodes (workers):**
  - The namenode also knows the **datanodes** on which all the blocks for a given file are located
  - it does not store block locations persistently, because this information is reconstructed from datanodes when the system starts.

➔ A **client** accesses the filesystem on behalf of the user by communicating with the namenode and datanodes.
  - The client presents a filesystem interface similar to a Portable Operating System Interface (POSIX), so the user code does not need to know about the namenode and datanodes to function.

➔ Datanodes are the workhorses of the filesystem.
  - They store and retrieve blocks when they are told to (by clients or the namenode), and they report back to the namenode periodically with lists of blocks that they are storing.
  - Without the namenode, the filesystem cannot be used.

➔ if the machine running the namenode were obliterated, all the files on the filesystem would be lost since there would be no way of knowing how to reconstruct the files from the blocks on the datanodes.
  - For this reason, it is important to make the namenode resilient to failure, and Hadoop provides two mechanisms for this.
  - **The first way is to back up the files** that make up the persistent state of the filesystem metadata. Hadoop can be configured so that the namenode writes its persistent state to multiple filesystems. These writes are synchronous and atomic. The usual configuration choice is to write to local disk as well as a remote NFS mount.
  - It is also possible to **run a secondary namenode**, which despite its name does not act as a namenode. Its main role is to periodically merge the namespace image with the edit log to prevent the edit log from becoming too large. The secondary namenode usually runs on a separate physical machine because it requires plenty of CPU and as much memory as the namenode to perform the

merge. It keeps a copy of the merged namespace image, which can be used in the event of the namenode failing. However, the state of the secondary namenode lags that of the primary, so in the event of total failure of the primary, data loss is almost certain. The usual course of action in this case is to copy the namenodes metadata files that are on NFS to the secondary and run it as the new primary

Block Caching

Normally a datanode reads blocks from disk, but for frequently accessed files the blocks may be explicitly cached in the datanode's memory, in an off-heap block cache. By default, a block is cached in only one datanode's memory, although the number is configurable on a per-file basis. Job schedulers (for MapReduce, Spark, and other frameworks) can take advantage of cached blocks by running tasks on the datanode where a block is cached, for increased read performance. A small lookup table used in a join is a good candidate for caching, for example. Users or applications instruct the namenode which files to cache (and for how long) by adding a cache directive to a cache pool. Cache pools are an administrative grouping for managing cache permissions and resource usage.

HDFS Federation

HDFS federation allows a cluster to scale by adding namenodes, each of which manages a portion of the filesystem namespace. The namenode keeps a reference to every file and block in the filesystem in memory, which means that on very large clusters with many files, memory becomes the limiting factor for scaling.

HDFS High Availability

The combination of replicating namenode metadata on multiple filesystems and using the secondary namenode to create checkpoints protects against data loss, but it does not provide high availability of the filesystem. The namenode is still a single point of failure (SPOF). If it did fail, all clients including MapReduce jobs would be unable to read, write, or list files, because the namenode is the sole repository of the metadata and the file-to-block mapping. In such an event, the whole Hadoop system would effectively be out of service until a new namenode could be brought online.

To recover from a failed namenode in this situation, an administrator starts a new primary namenode with one of the filesystem metadata replicas and configures datanodes and clients to use this new namenode. The new namenode is not able to serve requests until it has (i) loaded its namespace image into memory, (ii) replayed its edit log, and (iii) received enough block reports from the datanodes to leave safe mode.

Failover and fencing

The transition from the active namenode to the standby is managed by a new entity in the system called the **failover controller**. There are various failover controllers, but the default implementation uses ZooKeeper to ensure that only one namenode is active. Each namenode runs a lightweight failover controller process whose job it is to monitor its namenode for failures and trigger a failover should a namenode fail.

Failover may also be initiated manually by an administrator, for example, in the case of routine maintenance. This is known as a **graceful failover**, since the failover controller arranges an orderly transition for both namenodes to switch roles.

In the case of an ungraceful failover, however, it is impossible to be sure that the failed namenode has stopped running. For example, a slow network or a network partition can trigger a failover transition, even though the previously active namenode is still running and thinks it is still the active namenode. The HA implementation goes to great lengths to ensure that the previously active namenode is prevented from doing any damage and causing corruption a method known as **fencing**.

4.7.3 The Command-Line Interface

There are many other interfaces to HDFS, but the command line is one of the simplest and, to many developers, the most familiar. There are two properties that we set in the pseudo distributed configuration that deserve further explanation.

- The first is fs.defaultFS, set to hdfs://localhost/, which is used to set a default filesystem for Hadoop. Filesystems are specified by a URI, and here we have used an hdfs URI to configure Hadoop to use HDFS by default. The HDFS daemons will use this property to determine the host and port for the HDFS namenode. We'll be running it on localhost, on the default HDFS port, 8020. And HDFS clients will use this property to work out where the namenode is running so they can connect to it.
- We set the second property, dfs.replication, to 1 so that HDFS doesn't replicate filesystem blocks by the default factor of three. When running with a single datanode, HDFS can't replicate blocks to three datanodes, so it would perpetually warn about blocks being under-replicated. This setting solves that problem.

4.7.3.1 Basic Filesystem Operations

The filesystem is ready to be used, and we can do all of the usual filesystem operations, such as reading files, creating directories, moving files, deleting data, and listing directories. You can type hadoop fs -help to get detailed help on every command.

- Start by copying a file from the local filesystem to HDFS:
  *% hadoop fs -copyFromLocal input/docs/quangle.txt *
  hdfs://localhost/user/tom/quangle.txt

This command invokes Hadoop's filesystem shell command fs, which supports a number of subcommands in this case, we are running -copyFromLocal. The local file quangle.txt is copied to the file /user/tom/quangle.txt on the HDFS instance running on localhost. In fact, we could have omitted the scheme and host of the URI and picked up the default, hdfs://localhost, as specified in core-site.xml:

% hadoop fs -copyFromLocal input/docs/quangle.txt /user/tom/quangle.txt

- We also could have used a relative path and copied the file to our home directory in HDFS, which in this case is /user/tom:
  % hadoop fs -copyFromLocal input/docs/quangle.txt quangle.txt
- Let's copy the file back to the local filesystem and check whether it's the same:
  % hadoop fs -copyToLocal quangle.txt quangle.copy.txt
  % md5 input/docs/quangle.txt quangle.copy.txt

MD5 (input/docs/quangle.txt) = e7891a2627cf263a079fb0f18256ffb2
MD5 (quangle.copy.txt) = e7891a2627cf263a079fb0f18256ffb2

- Finally, let's look at an HDFS file listing. We create a directory first just to see how it is displayed in the listing:
  % hadoop fs -mkdir books
  % hadoop fs -ls .
  Found 2 items
  drwxr-xr-x - tom supergroup 0 2014-10-04 13:22 books
  -rw-r--r-- 1 tom supergroup 119 2014-10-04 13:21 quangle.txt

4.7.4 Hadoop Filesystems

Hadoop has an abstract notion of filesystems, of which HDFS is just one implementation. The Java abstract class org.apache.hadoop.fs. FileSystem represents the client interface to a filesystem in Hadoop, and there are several concrete implementations.

The main ones that ship with Hadoop are described in Figure 4.10.

Hadoop provides many interfaces to its filesystems, and it generally uses the URI scheme to pick the correct filesystem instance to communicate with. For example, the filesystem shell that we met in the previous section operates with all Hadoop filesystems. To list the files in the root directory of the local filesystem, type:

% hadoop fs -ls file:///

Although it is possible (and sometimes very convenient) to run MapReduce programs that access any of these filesystems, when you are processing large volumes of data you should choose a distributed filesystem that has the data locality optimization, notably HDFS.

Interfaces

Hadoop is written in Java, so most Hadoop filesystem interactions are mediated through the Java API. The filesystem shell, for example, is a Java application that uses the Java FileSystem class to provide filesystem operations. The other filesystem interfaces are discussed briefly in this section. These interfaces are most commonly used with HDFS, since the other filesystems in Hadoop typically have existing tools to access the underlying filesystem but many of them will work with any Hadoop filesystem.

Filesystem	URI scheme	Java implementation (all under org.apache.hadoop)	Description
Local	file	fs.LocalFileSystem	A filesystem for a locally connected disk with client-side checksums. Use RawLocalFileSystem for a local filesystem with no checksums. See "LocalFileSystem" on page 99.
HDFS	hdfs	hdfs.DistributedFileSystem	Hadoop's distributed filesystem. HDFS is designed to work efficiently in conjunction with MapReduce.
WebHDFS	webhdfs	hdfs.web.WebHdfsFileSystem	A filesystem providing authenticated read/write access to HDFS over HTTP. See "HTTP" on page 54.
Secure WebHDFS	swebhdfs	hdfs.web.SWebHdfsFileSystem	The HTTPS version of WebHDFS.
HAR	har	fs.HarFileSystem	A filesystem layered on another filesystem for archiving files. Hadoop Archives are used for packing lots of files in HDFS into a single archive file to reduce the namenode's memory usage. Use the hadoop archive command to create HAR files.
View	viewfs	viewfs.ViewFileSystem	A client-side mount table for other Hadoop filesystems. Commonly used to create mount points for federated namenodes (see "HDFS Federation" on page 48).
FTP	ftp	fs.ftp.FTPFileSystem	A filesystem backed by an FTP server.
S3	s3a	fs.s3a.S3AFileSystem	A filesystem backed by Amazon S3. Replaces the older s3n (S3 native) implementation.

Figure 4.10 Hadoop filesystems

HTTP

By exposing its filesystem interface as a Java API, Hadoop makes it awkward for non-Java applications to access HDFS. The HTTP REST API exposed by the WebHDFS protocol makes it easier for other languages to interact with HDFS. Note that the HTTP interface is slower than the native Java client, so should be avoided for very large data transfers if possible.

There are two ways of accessing HDFS over HTTP: directly, where the HDFS daemons serve HTTP requests to clients; and via a proxy (or proxies), which accesses HDFS on the client's behalf using the usual DistributedFileSystem API. The two ways are illustrated in Figure 4.11. Both use the WebHDFS protocol.

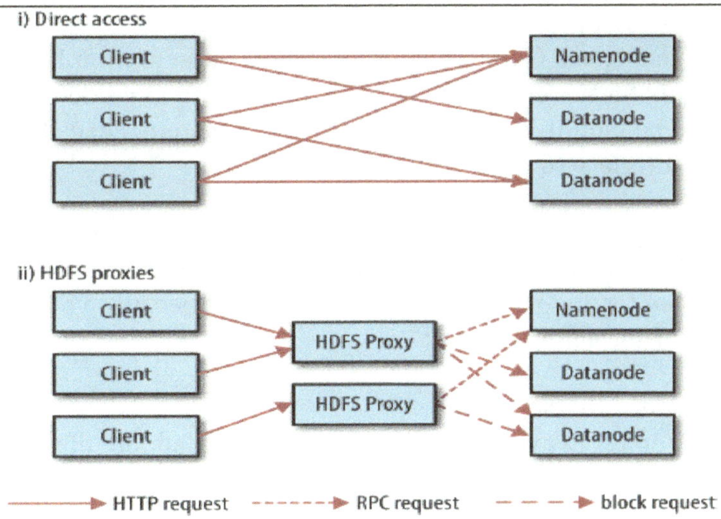

Figure 4.11 Accessing HDFS over HTTP directly and via a bank of HDFS proxies

In the first case, the embedded web servers in the namenode and datanodes act as WebHDFS endpoints. (WebHDFS is enabled by default, since dfs.webhdfs.enabled is set to true.) File metadata operations are handled by the namenode, while file read (and write) operations are sent first to the namenode, which sends an HTTP redirect to the client indicating the datanode to stream file data from (or to).

The second way of accessing HDFS over HTTP relies on one or more standalone proxy servers. (The proxies are stateless, so they can run behind a standard load balancer.) All traffic to the cluster passes through the proxy, so the client never accesses the namenode or datanode directly. This allows for stricter firewall and bandwidth-limiting policies to be put in place. It's common to use a proxy for transfers between Hadoop clusters located in different data centers, or when accessing a Hadoop cluster running in the cloud from an external network.

C

Hadoop provides a C library called libhdfs that mirrors the Java FileSystem interface. It works using the Java Native Interface (JNI) to call a Java filesystem client. There is also a libwebhdfs library that uses the WebHDFS interface. The C API is very similar to the Java one, but it typically lags the Java one, so some newer features may not be supported. You can find the header file, hdfs.h, in the include directory of the Apache Hadoop binary tarball distribution.

NFS

It is possible to mount HDFS on a local client's filesystem using Hadoop's NFSv3 gateway. You can then use Unix utilities (such as ls and cat) to interact with the filesystem, upload files, and in general use POSIX libraries to access the filesystem from any programming language. Appending to a file works, but random modifications of a file do not, since HDFS can only write to the end of a file.

FUSE

Filesystem in Userspace (FUSE) allows filesystems that are implemented in user space to be integrated as Unix filesystems. Hadoop's Fuse-DFS contrib module allows HDFS (or any Hadoop filesystem) to be mounted as a standard local filesystem. Fuse-DFS is implemented in C using libhdfs as the interface to HDFS.

4.7.5 The Java Interface

The Hadoop FileSystem class: the API for interacting with one of Hadoop's filesystems. This is very useful when testing your program, for example, because you can rapidly run tests using data stored on the local filesystem.

4.7.5.1 Reading Data from a Hadoop URL

One of the simplest ways to read a file from a Hadoop filesystem is by using a java.net.URL object to open a stream to read the data from. The general idiom is:

```
InputStream in = null;
try {
in = new URL("hdfs://host/path").openStream();
// process in
} finally {
IOUtils.closeStream(in);
}
```

By calling the setURLStreamHandlerFactory() method on URL with an instance of FsUrlStreamHandlerFactory is typically executed in a static block.

Example 4.1 shows a program for displaying files from Hadoop filesystems on standard output, like the Unix cat command.

Example 4.1 Displaying files from a Hadoop filesystem on standard output using a URLStreamHandler

```
public class URLCat {
static {
URL.setURLStreamHandlerFactory(new FsUrlStreamHandlerFactory());
}
public static void main(String[] args) throws Exception {
InputStream in = null;
try {
in = new URL(args[0]).openStream();
IOUtils.copyBytes(in, System.out, 4096, false);
} finally { IOUtils.closeStream(in);}}}
```

sample run:7

```
% export HADOOP_CLASSPATH=hadoop-examples.jar
% hadoop URLCat hdfs://localhost/user/tom/quangle.txt
On the top of the Crumpetty Tree
The Quangle Wangle sat,
```

But his face you could not see,
On account of his Beaver Hat.

We make use of the handy IOUtils class that comes with Hadoop for closing the stream in the finally clause, and also for copying bytes between the input stream and the output stream. The last two arguments to the copyBytes() method are the buffer size used for copying and whether to close the streams when the copy is complete.

4.7.5.2 Reading Data Using the FileSystem API

sometimes it is impossible to set a URLStreamHand lerFactory for your application. A file in a Hadoop filesystem is represented by a Hadoop Path object. You can think of a Path as a Hadoop filesystem URI, such as hdfs://localhost/user/tom/quangle.txt.

FileSystem is a general filesystem API, so the first step is to

retrieve an instance for the filesystem

public static FileSystem get(Configuration conf) throws IOException
public static FileSystem get(URI uri, Configuration conf) throws IOException
public static FileSystem get(URI uri, Configuration conf, String user)
throws IOException

A Configuration object encapsulates a client or server's configuration, which is set using configuration files read from the classpath, such as etc/hadoop/core-site.xml.

- The first method returns the default filesystem (as specified in core-site.xml, or the default local filesystem if not specified there).
- The second uses the given URI's scheme and authority to determine the filesystem to use, falling back to the default filesystem if no scheme is specified in the given URI.
- The third retrieves the filesystem as the given user, which is important in the context of security.

retrieve a local file system instance

Use the convenience method getLocal():

public static LocalFileSystem getLocal(Configuration conf) throws IOException

With a FileSystem instance in hand, we invoke an open() method to get the input stream for a file:

public FSDataInputStream open(Path f) throws IOException

public abstract FSDataInputStream open(Path f, int bufferSize) throws IOException

Example 4.2. Displaying files from a Hadoop filesystem on standard output by using the FileSystem directly

```
public class FileSystemCat {
public static void main(String[] args) throws Exception {
String uri = args[0];
Configuration conf = new Configuration();
```

```
FileSystem fs = FileSystem.get(URI.create(uri), conf);
InputStream in = null;
try {
in = fs.open(new Path(uri));
IOUtils.copyBytes(in, System.out, 4096, false);
} finally {
IOUtils.closeStream(in);
}
}
}
```

The program runs as follows:

```
% hadoop FileSystemCat hdfs://localhost/user/tom/quangle.txt
On the top of the Crumpetty Tree
The Quangle Wangle sat,
But his face you could not see,
On account of his Beaver Hat.
```

FSDataInputStream

The open() method on FileSystem actually returns an FSDataInputStream rather than a standard java.io class. This class is a specialization of java.io.DataInputStream with support for random access, so you can read from any part of the stream:

```
package org.apache.hadoop.fs;
public class FSDataInputStream extends DataInputStream
implements Seekable, PositionedReadable {
// implementation elided
}
```

The Seekable interface permits seeking to a position in the file and provides a query method for the current offset from the start of the file (getPos()):

```
public interface Seekable {
void seek(long pos) throws IOException;
long getPos() throws IOException;
}
```

Calling seek() with a position that is greater than the length of the file will result in an IOException. Unlike the skip() method of java.io.InputStream, which positions the stream at a point later than the current position, seek() can move to an arbitrary, absolute position in the file.

Example 4.3 Displaying files from a Hadoop filesystem on standard output twice, by using seek()

```
public class FileSystemDoubleCat {
```

```
public static void main(String[] args) throws Exception {
String uri = args[0];
Configuration conf = new Configuration();
FileSystem fs = FileSystem.get(URI.create(uri), conf);
FSDataInputStream in = null;
try {
in = fs.open(new Path(uri));
IOUtils.copyBytes(in, System.out, 4096, false);
in.seek(0); // go back to the start of the file
IOUtils.copyBytes(in, System.out, 4096, false);
} finally {
IOUtils.closeStream(in);
}
}
}
Here's the result of running it on a small file:
% hadoop FileSystemDoubleCat hdfs://localhost/user/tom/quangle.txt
On the top of the Crumpetty Tree
The Quangle Wangle sat,
But his face you could not see,
On account of his Beaver Hat.
On the top of the Crumpetty Tree
The Quangle Wangle sat,
But his face you could not see,
On account of his Beaver Hat.
```

FSDataInputStream also implements the PositionedReadable interface for reading parts of a file at a given offset:

```
public interface PositionedReadable {
public int read(long position, byte[] buffer, int offset, int length)
throws IOException;
public void readFully(long position, byte[] buffer, int offset, int length)
throws IOException;
public void readFully(long position, byte[] buffer) throws IOException;
}
```

The read() method reads up to length bytes from the given position in the file into the buffer at the given offset in the buffer. The return value is the number of bytes actually read; callers should check this value, as it may be less than length.

4.7.5.3 Writing Data

The FileSystem class has a number of methods for creating a file. The simplest is the method that takes a Path object for the file to be created and returns an output stream to write to:

```
public FSDataOutputStream create(Path f) throws IOException
```

There are overloaded versions of this method that allow you to specify whether to forcibly overwrite existing files, the replication factor of the file, the buffer size to use when writing the file, the block size for the file, and file permissions. There's also an overloaded method for passing a callback interface, Progressable

```
package org.apache.hadoop.util;
public interface Progressable {
public void progress();
}
```

As an alternative to creating a new file, you can append to an existing file using the append() method (there are also some other overloaded versions):

```
public FSDataOutputStream append(Path f) throws IOException
```

Example 4.4 Copying a local file to a Hadoop filesystem

```
public class FileCopyWithProgress {
public static void main(String[] args) throws Exception {
String localSrc = args[0];
String dst = args[1];
InputStream in = new BufferedInputStream(new FileInputStream(localSrc));
Configuration conf = new Configuration();
FileSystem fs = FileSystem.get(URI.create(dst), conf);
OutputStream out = fs.create(new Path(dst), new Progressable() {
public void progress() {
System.out.print(".");
}
});
IOUtils.copyBytes(in, out, 4096, true);
}
}
```

Typical usage:

```
% hadoop FileCopyWithProgress input/docs/1400-8.txt
hdfs://localhost/user/tom/1400-8.txt
```

FSDataOutputStream

The create() method on FileSystem returns an FSDataOutputStream, which, like FSDataInputStream, has a method for querying the current position in the file:

```
package org.apache.hadoop.fs;
public class FSDataOutputStream extends DataOutputStream implements Syncable {
public long getPos() throws IOException {
// implementation elided
}
// implementation elided
```

```
}
```

However, unlike FSDataInputStream, FSDataOutputStream does not permit seeking. This is because HDFS allows only sequential writes to an open file or appends to an already written file.

4.7.5.4 Directories

FileSystem provides a method to create a directory:

public boolean mkdirs(Path f) throws IOException

This method creates all of the necessary parent directories if they don't already exist, just like the java.io.File's mkdirs() method. It returns true if the directory (and all parent directories) was (were) successfully created. Often, don't need to explicitly create a directory, because writing a file by calling create() will automatically create any parent directories.

4.7.5.5 Querying the Filesystem

File metadata: FileStatus

An important feature of any filesystem is the ability to navigate its directory structure and retrieve information about the files and directories that it stores. The FileStatus class encapsulates filesystem metadata for files and directories, including file length, block size, replication, modification time, ownership, and permission information.

The method getFileStatus() on FileSystem provides a way of getting a FileStatus object for a single file or directory. Example 4.5 shows an example of its use.

Example 4.5. Demonstrating file status information

```
public class ShowFileStatusTest {
private MiniDFSCluster cluster; // use an in-process HDFS cluster for testing
private FileSystem fs;
@Before
public void setUp() throws IOException {
Configuration conf = new Configuration();
if (System.getProperty("test.build.data") == null) {
System.setProperty("test.build.data", "/tmp");
}
cluster = new MiniDFSCluster.Builder(conf).build();
fs = cluster.getFileSystem();
OutputStream out = fs.create(new Path("/dir/file"));
out.write("content".getBytes("UTF-8"));
out.close();
}
@After
public void tearDown() throws IOException {
if (fs != null) { fs.close(); }
if (cluster != null) { cluster.shutdown(); }
}
```

```
@Test(expected = FileNotFoundException.class)
public void throwsFileNotFoundForNonExistentFile() throws IOException {
fs.getFileStatus(new Path("no-such-file"));
}
@Test
public void fileStatusForFile() throws IOException {
Path file = new Path("/dir/file");
FileStatus stat = fs.getFileStatus(file);
assertThat(stat.getPath().toUri().getPath(), is("/dir/file"));
assertThat(stat.isDirectory(), is(false));
assertThat(stat.getLen(), is(7L));
assertThat(stat.getModificationTime(),
is(lessThanOrEqualTo(System.currentTimeMillis())));
assertThat(stat.getReplication(), is((short) 1));
assertThat(stat.getBlockSize(), is(128 * 1024 * 1024L));
assertThat(stat.getOwner(), is(System.getProperty("user.name")));
assertThat(stat.getGroup(), is("supergroup"));
assertThat(stat.getPermission().toString(), is("rw-r--r--"));
}
@Test
public void fileStatusForDirectory() throws IOException {
Path dir = new Path("/dir");
FileStatus stat = fs.getFileStatus(dir);
assertThat(stat.getPath().toUri().getPath(), is("/dir"));
assertThat(stat.isDirectory(), is(true));
assertThat(stat.getLen(), is(0L));
assertThat(stat.getModificationTime(),
is(lessThanOrEqualTo(System.currentTimeMillis())));
assertThat(stat.getReplication(), is((short) 0));
assertThat(stat.getBlockSize(), is(0L));
assertThat(stat.getOwner(), is(System.getProperty("user.name")));
assertThat(stat.getGroup(), is("supergroup"));
assertThat(stat.getPermission().toString(), is("rwxr-xr-x"));
}
}
```

If no file or directory exists, a FileNotFoundException is thrown. However, if you are interested only in the existence of a file or directory, the exists() method on FileSystem is more convenient:

```
public boolean exists(Path f) throws IOException
```

Listing files

Finding information on a single file or directory is useful, but you also often need to be able to list the contents of a directory. That's what FileSystem's listStatus() methods are for:

```
public FileStatus[] listStatus(Path f) throws IOException
```

```
public FileStatus[] listStatus(Path f, PathFilter filter) throws IOException
public FileStatus[] listStatus(Path[] files) throws IOException
public FileStatus[] listStatus(Path[] files, PathFilter filter)
throws IOException
```

When the argument is a file, the simplest variant returns an array of FileStatus objects of length 1. When the argument is a directory, it returns zero or more FileStatus objects representing the files and directories contained in the directory.

Example 4.6 Showing the file statuses for a collection of paths in a Hadoop filesystem

```
public class ListStatus {
public static void main(String[] args) throws Exception {
String uri = args[0];
Configuration conf = new Configuration();
FileSystem fs = FileSystem.get(URI.create(uri), conf);
Path[] paths = new Path[args.length];
for (int i = 0; i < paths.length; i++) {
paths[i] = new Path(args[i]);
}
FileStatus[] status = fs.listStatus(paths);
Path[] listedPaths = FileUtil.stat2Paths(status);
for (Path p : listedPaths) {
System.out.println(p);
}
}
}
```

We can use this program to find the union of directory listings for a collection of paths:

```
% hadoop ListStatus hdfs://localhost/ hdfs://localhost/user/tom
hdfs://localhost/user
hdfs://localhost/user/tom/books
hdfs://localhost/user/tom/quangle.txt
```

File patterns

It is a common requirement to process sets of files in a single operation. For example, a MapReduce job for log processing might analyze a month's worth of files contained in a number of directories. Rather than having to enumerate each file and directory to specify the input, it is convenient to use wildcard characters to match multiple files with a single expression, an operation that is known as globbing. Hadoop provides two FileSystem methods for processing globs:

```
public FileStatus[] globStatus(Path pathPattern) throws IOException
public FileStatus[] globStatus(Path pathPattern, PathFilter filter)
throws IOException
```

The globStatus() methods return an array of FileStatus objects whose paths match the supplied pattern, sorted by path. An optional PathFilter can be specified to restrict the matches further.

Imagine that logfiles are stored in a directory structure organized hierarchically by date. So, logfiles for the last day of 2007 would go in a directory named /2007/12/31, for example. Suppose that the full file listing is:

```
/
├── 2007/
│   └── 12/
│   ├── 30/
│   └── 31/
└── 2008/
└── 01/
├── 01/
└── 02/
```

Here are some file globs and their expansions:

Glob	Expansion
/*	/2007 /2008
/*/*	/2007/12 /2008/01
/*/12/*	/2007/12/30 /2007/12/31
/200?	/2007 /2008
/200[78]	/2007 /2008
/200[7-8]	/2007 /2008
/200[^01234569]	/2007 /2008
/*/*/{31,01}	/2007/12/31 /2008/01/01
/*/*/3{0,1}	/2007/12/30 /2007/12/31
/*/{12/31,01/01}	/2007/12/31 /2008/01/01

PathFilter

Glob patterns are not always powerful enough to describe a set of files you want to access. For example, it is not generally possible to exclude a particular file using a glob pattern. The listStatus() and globStatus() methods of FileSystem take an optional PathFilter, which allows programmatic control over matching:

```
package org.apache.hadoop.fs;
public interface PathFilter {
boolean accept(Path path);
}
```

PathFilter is the equivalent of java.io.FileFilter for Path objects rather than File objects.

Example 4.7 A PathFilter for excluding paths that match a regular expression

```
public class RegexExcludePathFilter implements PathFilter {
private final String regex;
```

```
public RegexExcludePathFilter(String regex) {
this.regex = regex;
}
public boolean accept(Path path) {
return !path.toString().matches(regex);
}
}
```

The filter passes only those files that don't match the regular expression. After the glob picks out an initial set of files to include, the filter is used to refine the results. For example:

fs.globStatus(new Path("/2007/*/*"), new RegexExcludeFilter("^.*/2007/12/31$")) will expand to /2007/12/30.

4.7.6 Deleting Data

Use the delete() method on FileSystem to permanently remove files or directories:

public boolean delete(Path f, boolean recursive) throws IOException

If f is a file or an empty directory, the value of recursive is ignored. A nonempty directory is deleted, along with its contents, only if recursive is true (otherwise, an IOException is thrown).

Data Flow

Anatomy of a File Read

To get an idea of how data flows between the client interacting with HDFS, the namenode, and the datanodes, consider Figure 4.12, which shows the main sequence of events when reading a file.

- The client opens the file it wishes to read by calling open() on the FileSystem object, which for HDFS is an instance of DistributedFileSystem (step 1 in Figure 4.12).
- DistributedFileSystem calls the namenode, using remote procedure calls (RPCs), to determine the locations of the first few blocks in the file (step 2).
- For each block, the namenode returns the addresses of the datanodes that have a copy of that block. Furthermore, the datanodes are sorted according to their proximity to the client.
- If the client is itself a datanode (in the case of a MapReduce task, for instance), the client will read from the local datanode if that datanode hosts a copy of the block.
- The DistributedFileSystem returns an FSDataInputStream to the client for it to read data from. FSDataInputStream in turn wraps a DFSInputStream, which manages the datanode and namenode I/O. The client then calls read() on the stream (step 3).
- DFSInputStream, which has stored the datanode addresses for the first few blocks in the file, then connects to the first(closest) datanode for the first block in the file. Data is streamed from the datanode back to the client, which calls read() repeatedly on the stream (step 4).

- When the end of the block is reached, DFSInputStream will close the connection to the datanode, then find the best datanode for the next block (step 5).
- Blocks are read in order, with the DFSInputStream opening new connections to datanodes as the client reads through the stream. It will also call the namenode to retrieve the datanode locations for the next batch of blocks as needed. When the client has finished reading, it calls close() on the FSDataInputStream (step 6).

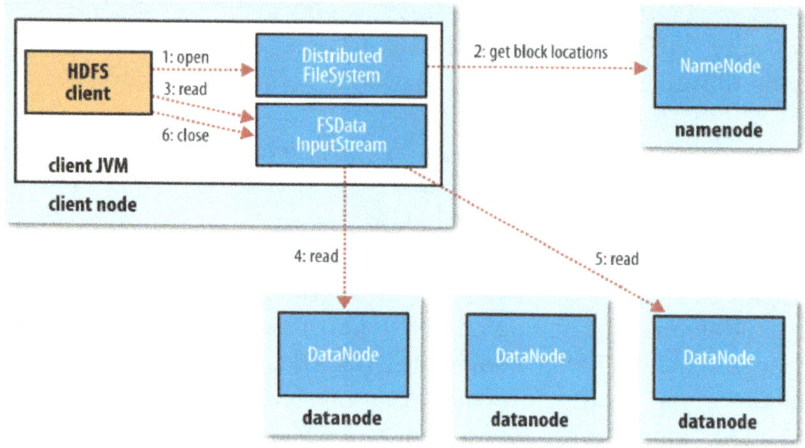

Figure 4.12. A client reading data from HDFS

Anatomy of a File Write

Next we'll look at how files are written to HDFS. Although quite detailed, it is instructive to understand the data flow because it clarifies HDFS's coherency model.

We're going to consider the case of creating a new file, writing data to it, then closing the file. This is illustrated in Figure 4.13.

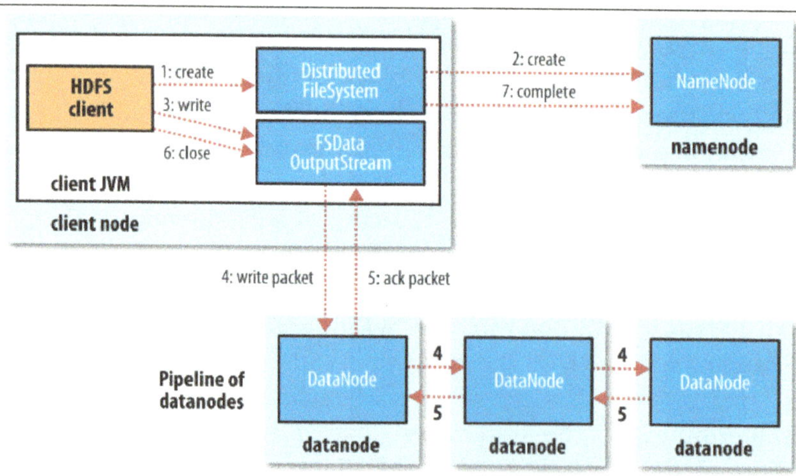

Figure 4.13 A client writing data to HDFS

- The client creates the file by calling create() on DistributedFileSystem (step 1 in Figure 4.13).
- DistributedFileSystem makes an RPC call to the namenode to create a new file in the filesystem's namespace, with no blocks associated with it (step 2).
- The namenode performs various checks to make sure the file doesn't already exist and that the client has the right permissions to create the file. If these checks pass, the namenode makes a record of the new file; otherwise, file creation fails and the client is thrown an IOException. The DistributedFileSystem returns an FSDataOutputStream for the client to start writing data to. Just as in the read case, FSDataOutputStream wraps a DFSOutputStream, which handles communication with the datanodes and namenode.
- As the client writes data (step 3), the DFSOutputStream splits it into packets, which it writes to an internal queue called the data queue. The data queue is consumed by the DataStreamer, which is responsible for asking the namenode to allocate new blocks by picking a list of suitable datanodes to store the replicas. The list of datanodes forms a pipeline, and here we'll assume the replication level is three, so there are three nodes in the pipeline.
- The DataStreamer streams the packets to the first datanode in the pipeline, which stores each packet and forwards it to the second datanode in the pipeline. Similarly, the second datanode stores the packet and forwards it to the third (and last) datanode in the pipeline (step 4).
- The DFSOutputStream also maintains an internal queue of packets that are waiting to be acknowledged by datanodes, called the ack queue. A packet is removed from the ack queue only when it has been acknowledged by all the datanodes in the pipeline (step 5).
- When the client has finished writing data, it calls close() on the stream (step 6).
- This action flushes all the remaining packets to the datanode pipeline and waits for acknowledgments before contacting the namenode to signal that the file is complete (step 7).

Coherency Model

A coherency model for a filesystem describes the data visibility of reads and writes for a file. HDFS trades off some POSIX requirements for performance, so some operations may behave differently than you expect them to.

After creating a file, it is visible in the filesystem namespace, as expected:

```
Path p = new Path("p");
fs.create(p);
assertThat(fs.exists(p), is(true));
```

However, any content written to the file is not guaranteed to be visible, even if the stream is flushed. So, the file appears to have a length of zero:

```
Path p = new Path("p");
OutputStream out = fs.create(p);
out.write("content".getBytes("UTF-8"));
out.flush();
```

```
assertThat(fs.getFileStatus(p).getLen(), is(0L));
```

Once more than a block's worth of data has been written, the first block will be visible to new readers. This is true of subsequent blocks, too: it is always the current block being written that is not visible to other readers.

HDFS provides a way to force all buffers to be flushed to the datanodes via the hflush() method on FSDataOutputStream. After a successful return from hflush(), HDFS guarantees that the data written up to that point in the file has reached all the datanodes in the write pipeline and is visible to all new readers:

```
Path p = new Path("p");
FSDataOutputStream out = fs.create(p);
out.write("content".getBytes("UTF-8"));
out.hflush();
assertThat(fs.getFileStatus(p).getLen(), is(((long) "content".length())));
```

The behavior of hsync() is similar to that of the fsync() system call in POSIX that commits buffered data for a file descriptor. For example, using the standard Java API to write a local file, we are guaranteed to see the content after flushing the stream and synchronizing:

```
FileOutputStream out = new FileOutputStream(localFile);
out.write("content".getBytes("UTF-8"));
out.flush(); // flush to operating system
out.getFD().sync(); // sync to disk
assertThat(localFile.length(), is(((long) "content".length())));
```

Closing a file in HDFS performs an implicit hflush(), too:

```
Path p = new Path("p");
OutputStream out = fs.create(p);
out.write("content".getBytes("UTF-8"));
out.close();
assertThat(fs.getFileStatus(p).getLen(), is(((long) "content".length())));
```

Consequences for application design

This coherency model has implications for the way you design applications. With no calls to hflush() or hsync(), you should be prepared to lose up to a block of data in the event of client or system failure. For many applications, this is unacceptable, so you should call hflush() at suitable points, such as after writing a certain number of records or number of bytes. Though the hflush() operation is designed to not unduly tax HDFS,it does have some overhead (and hsync() has more), so there is a trade-off between data robustness and throughput. What constitutes an acceptable trade-off is application dependent, and suitable values can be selected after measuring your application's performance with different hflush() (or hsync()) frequencies.

4.7.7 Parallel Copying with distcp

The HDFS access patterns focus on single-threaded access. It's possible to act on a collection of files by specifying file globs, for example but for efficient parallel processing of these files would have to write a program yourself. Hadoop comes with a useful program called distcp for copying data to and from Hadoop filesystems in parallel.

One use for distcp is as an efficient replacement for hadoop fs -cp. For example, you can copy one file to another with:10

% hadoop distcp file1 file2

You can also copy directories:

% hadoop distcp dir1 dir2

If dir2 does not exist, it will be created, and the contents of the dir1 directory will be copied there. You can specify multiple source paths, and all will be copied to the destination. If dir2 already exists, then dir1 will be copied under it, creating the directory structure dir2/dir1. If this isn't what you want, you can supply the -overwrite option to keep the same directory structure and force files to be overwritten. You can also update only the files that have changed using the -update option. This is best shown with an example.

If we changed a file in the dir1 subtree, we could synchronize the change with dir2 by running:

% hadoop distcp -update dir1 dir2

distcp is implemented as a MapReduce job where the work of copying is done by the maps that run in parallel across the cluster. There are no reducers. Each file is copied by a single map, and distcp tries to give each map approximately the same amount of data by bucketing files into roughly equal allocations. By default, up to 20 maps are used, but this can be changed by specifying the -m argument to distcp.

A very common use case for distcp is for transferring data between two HDFS clusters. For example, the following creates a backup of the first cluster's /foo directory on the second:

% hadoop distcp -update -delete -p hdfs://namenode1/foo hdfs://namenode2/foo

The -delete flag causes distcp to delete any files or directories from the destination that are not present in the source, and -p means that file status attributes like permissions, block size, and replication are preserved. You can run distcp with no arguments to see precise usage instructions. If the two clusters are running incompatible versions of HDFS, then you can use the webhdfs protocol to distcp between them:

% hadoop distcp webhdfs://namenode1:50070/foo webhdfs://namenode2:50070/foo

Another variant is to use an HttpFs proxy as the distcp source or destination (again using the webhdfs protocol), which has the advantage of being able to set firewall and bandwidth controls.

Keeping an HDFS Cluster Balanced

When copying data into HDFS, it's important to consider cluster balance. HDFS works best when the file blocks are evenly spread across the cluster, so you want to ensure that distcp doesn't disrupt this. For example, if you specified -m 1, a single map would do the copy, which apart from being slow and not using the cluster resources efficiently would mean that the first replica of each block would reside on the node running the map (until the disk filled up). The second and third replicas would be spread across the cluster, but this one node would be unbalanced. By having more maps than nodes in the cluster, this problem is avoided. For this reason, it's best to start by running distcp with the default of 20 maps per node. However, it's not always possible to prevent a cluster from becoming unbalanced. Perhaps you want to limit the number of maps so that some of the nodes can be used by other jobs. In this case, you can use the balancer tool to subsequently even out the block distribution across the cluster.

5. SECURITY

Trust models for Grid security environment – Authentication and Authorization methods – Grid security infrastructure – Cloud Infrastructure security: network, host and application level – aspects of data security, provider data and its security, Identity and access management architecture, IAM practices in the cloud, SaaS, PaaS, IaaS availability in the cloud, Key privacy issues in the cloud.

5. Grid Application Trends and Security Measures

Users and resources from multiple administrative organizations are involved in a grid. We need to prevent not only network attacks to data and resources, but also combat the selfishness of grid users and remove the distrust among users. More importantly, we should be aware of the security challenges in grid applications. We need to establish a trust mode for grid computing. The principles of some grid security enforcement schemes and mechanisms are presented. Specifically, we describe methods to cope with the problems of user authentication and resource-access authorization.

5.1 Trust Models for Grid Security Enforcement

Many potential security issues may occur in a grid environment if qualified security mechanisms are not in place. These issues include

- network sniffers
- out-of-control access
- faulty operation
- malicious operation
- integration of local security mechanisms
- delegation
- dynamic resources and services
- attack provenance, and so on.

Computational grids are motivated by the **desire to share processing resources among many organizations to solve large-scale problems**. Indeed, grid sites may exhibit unacceptable security conditions and system vulnerabilities.

Security-assurance condition

- a user job demands the resource site to provide security assurance by issuing a **security demand** (SD).
- The site needs to reveal its trustworthiness, called its **trust index** (TI).
- These two parameters must satisfy a **security-assurance condition**: TI ≥ SD during the job mapping process.

When determining its security demand, users usually care about some typical attributes. These attributes and their values are dynamically changing and depend heavily on the trust model, security policy, accumulated reputation, self-defense capability, attack history, and site vulnerability.

Three challenges are outlined below to establish the trust among grid sites.

[1]. the first challenge is integration with existing systems and technologies.

- The resources sites in a grid are usually heterogeneous and autonomous.
- It is unrealistic to expect that a single type of security can be compatible with and adopted by every hosting environment.
- At the same time, existing security infrastructure on the sites cannot be replaced overnight.
- Thus, to be successful, grid security architecture needs to step up to the challenge of integrating with existing security architecture and models across platforms and hosting environments.

[2]. the second challenge is interoperability with different "hosting environments."

- Services are often invoked across multiple domains, and need to be able to interact with one another.
- The interoperation is demanded at the protocol, policy, and identity levels.
- For all these levels, interoperation must be protected securely.

[3]. the third challenge is to construct trust relationships among interacting hosting environments.

- Grid service requests can be handled by combining resources on multiple security domains.
- Trust relationships are required by these domains during the end-to-end traversals.
- A service needs to be open to friendly and interested entities so that they can submit requests and access securely.

Resource sharing among entities is one of the major goals of grid computing.

- A trust relationship must be established before the entities in the grid interoperate with one another.
- The entities have to choose other entities that can meet the requirements of trust to coordinate with.
- The entities that submit requests should believe the resource providers will try to process their requests and return the results with a specified QoS.
- To create the proper trust relationship between grid entities, **two kinds of trust models** are often used.
    - ➔ One is the **PKI-based model**, which mainly exploits the PKI to authenticate and authorize entities
    - ➔ The other is the **reputation-based model**.

The grid aims to construct a large-scale network computing system by integrating distributed, heterogeneous, and autonomous resources. The security challenges faced by the grid are much greater than other computing systems. Before any effective sharing and cooperation occurs, a trust relationship has to be established among participants. Otherwise, not only will participants be reluctant to share their resources and services, but also the grid may cause a lot of damage.

5.1.1 A Generalized Trust Model

Figure 5.1 shows a general trust model. At the bottom, we identify three major factors which influence the trustworthiness of a resource site.

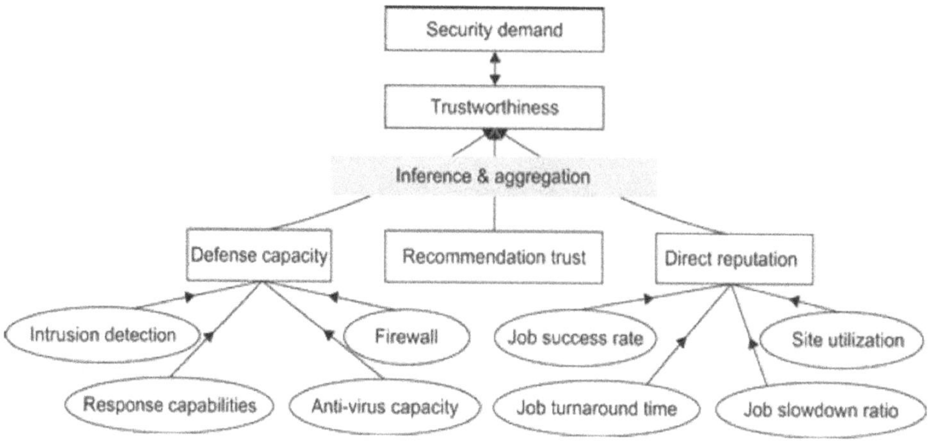

Figure 5.1 A general trust model for grid computing

An **inference module** is required to aggregate these factors. Followings are some existing inference or aggregation methods.

- ➔ An **intra-site fuzzy inference procedure** is called to assess defense capability and direct reputation.
- ➔ **Defense capability** is decided by the firewall, intrusion detection system (IDS), intrusion response capability, and anti-virus capacity of the individual resource site.
- ➔ **Direct reputation** is decided based on the job success rate, site utilization, job turnaround time, and job slowdown ratio measured.

Recommended trust is also known as secondary trust and is obtained indirectly over the grid network.

5.1.2 Reputation-Based Trust Model

- In a reputation-based model, jobs are sent to a resource site only when the site is trustworthy to meet users' demands. The site trustworthiness is usually calculated from the following information:
    - ➔ the defense capability
    - ➔ direct reputation
    - ➔ recommendation trust
- The **defense capability** refers to the site's ability to protect itself from danger. It is assessed according to such factors as intrusion detection, firewall, response capabilities, anti-virus capacity, and so on.
- **Direct reputation** is based on experiences of prior jobs previously submitted to the site.

- The **reputation** is measured by many factors such as prior job execution success rate, cumulative site utilization, job turnaround time, job slowdown ratio, and so on.
- A **positive experience** associated with a site will improve its reputation. On the contrary, a **negative experience** with a site will decrease its reputation.

5.1.2 A Fuzzy-Trust Model

- In this model the job security demand (SD) is supplied by the user programs.
- The trust index (TI) of a resource site is aggregated through the fuzzy-logic inference process over all related parameters.
- One can use a **two-level fuzzy logic** to estimate the aggregation of numerous trust parameters and security attributes into scalar quantities that are easy to use in the job scheduling and resource mapping process.
- The TI is normalized as a single real number with
    - 0 representing the condition with the highest risk at a site
    - 1 representing the condition which is totally risk-free or fully trusted.
- The fuzzy inference is accomplished through four steps:
    - fuzzification,
    - inference,
    - aggregation, and
    - defuzzification.
- The second salient feature of the trust model is that if a site's trust index cannot match the job security demand (i.e., SD > TI), the trust model could deduce detailed security features to guide the site security upgrade as a result of tuning the fuzzy system.

5.2 Authentication and Authorization Methods

The major authentication methods in the grid include passwords, PKI, and Kerberos.

- The password is the simplest method to identify users, but the most vulnerable one to use.
- The PKI is the most popular method supported by GSI.
    - To implement PKI, we use a trusted third party, called the certificate authority (CA).
    - Each user applies a unique pair of public and private keys.
    - The public keys are issued by the CA by issuing a certificate, after recognizing a legitimate user.
    - The private key is exclusive for each user to use, and is unknown to any other users.
    - A digital certificate in IEEE X.509 format consists of the user name, user public key, CA name, and a secrete signature of the user.

The following example illustrates the use of a PKI service in a grid environment.

Example: Trust Delegation Using the Proxy Credential in GSI

The PKI is not strong enough for user authentication in a grid. Figure 5.2 shows a scenario where a sequence of trust delegation is necessary. Bob and Charlie both trust Alice, but Charlie does not trust Bob. Now, Alice submits a task Z to Bob. The task Z demands many resources for Bob to use, independently. Bob forwards a subtask Y of Z to Charlie. Because

Charlie does not trust Bob and is not sure whether Y is really originally requested by Alice, the subtask Y from Bob is rejected for resources by
Charlie.

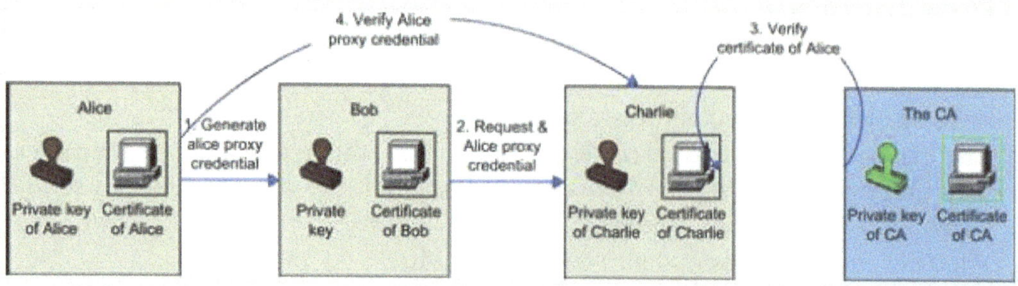

Figure 5.2 Interactions among multiple parties in a sequence of trust delegation operations using the PKI services in a GT4-enabled grid environment.

For Charlie to accept the subtask Y, Bob needs to show Charlie some proof of entrust from Alice. A proxy credential is the solution proposed by GSI. A proxy credential is a temporary certificate generated by a user. Two benefits are seen by using proxy credentials. First, the proxy credential is used by its holder to act on behalf of the original user or the delegating party. A user can temporarily delegate his right to a proxy. Second, single sign-on can be achieved with a sequence of credentials passed along the trust chain.

The delegating party (Alice) need not verify the remote intermediate parties in a trust chain. The only difference between the proxy credential and a digital certificate is that the proxy credential is not signed by a CA. We need to known the relationship among the certificates of the CA and Alice, and proxy credential of Alice. The CA certificate is signed first with its own private key. Second, the certificate Alice holds is signed with the private key of the CA. Finally, the proxy credential sent to her proxy (Bob) is signed with her private key. The procedure delegates the rights of Alice to Bob by using the proxy credential.

First, the generation of the proxy credential is similar to the procedure of generating a user certificate in the traditional PKI. Second, when Bob acts on behalf of Alice, he sends the request together with Alice's proxy credential and the Alice certificate to Charlie. Third, after obtaining the proxy credential, Charlie finds out that the proxy credential is signed by Alice. So he tries to verify the identity of Alice and finds Alice trustable. Finally, Charlie accepts Bob's requests on behalf of Alice. This is called a trust delegation chain.

5.2.1 Authorization for Access Control

The authorization is a process to exercise access control of shared resources. Decisions can be made either at the access point of service or at a centralized place. Typically, the resource is a host that provides processors and storage for services deployed on it. Based on a set predefined policies or rules, the resource may enforce access for local services. The central authority is a special entity which is capable of issuing and revoking polices of access rights granted to remote accesses. The authority can be classified into three categories: attribute authorities, policy authorities, and identity authorities. Attribute authorities issue attribute assertions; policy authorities issue authorization policies; identity authorities issue certificates. The authorization server makes the final authorization decision.

5.2.2 Three Authorization Models

Figure 5.3 shows three authorization models. The subject is the user and the resource refers to the machine side.

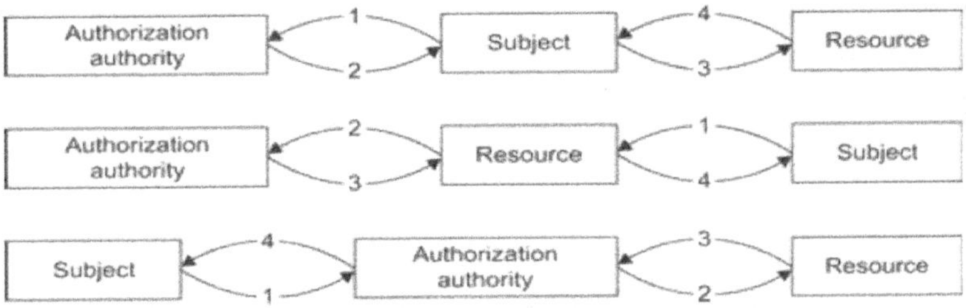

Figure 5.3 Three authorization models: the subject-push model, resource-pulling model, and the authorization agent model.

- The **subject-push model** is shown at the top diagram. The user conducts handshake with the authority first and then with the resource site in a sequence.
- The **resource-pulling model** puts the resource in the middle. The user checks the resource first. Then the resource contacts its authority to verify the request, and the authority authorizes at step 3. Finally the resource accepts or rejects the request from the subject at step 4.
- The **authorization agent model** puts the authority in the middle. The subject check with the authority at step 1 and the authority makes decisions on the access of the requested resources. The authorization process is complete at steps 3 and 4 in the reverse direction.

5.3 Grid Security Infrastructure (GSI)

The grid is increasingly deployed as a common approach to constructing dynamic, inter domain, distributed computing and data collaborations, "**lack of security/trust between different services**" is still an important challenge of the grid.

The grid requires a security infrastructure with the following properties:

- easy to use
- conforms with the VO's security needs while working well with site policies of each resource provider site
- and provides appropriate authentication and encryption of all interactions

Grid Security Infrastructure

- The GSI is an important step toward satisfying these requirements.

- As a well-known security solution in the grid environment, GSI is a portion of the Globus Toolkit and provides fundamental security services needed to support grids, including supporting for
  - ➔ message protection
  - ➔ authentication and delegation
  - ➔ and authorization
- GSI enables secure authentication and communication over an open network, and permits mutual authentication across and among distributed sites with single sign-on capability.
- No centrally managed security system is required, and the grid maintains the integrity of its members' local policies.
- GSI supports both message-level security, which supports the WS Security standard and the WS-Secure Conversation specification to provide message protection for SOAP messages, and transport-level security, which means authentication via TLS with support for X.509 proxy certificates.

5.3.1 GSI Functional Layers

GT4 provides distinct WS and pre-WS authentication and authorization capabilities. Both build on the same base, namely the X.509 standard and entity certificates and proxy certificates, which are used to identify persistent entities such as users and servers and to support the temporary delegation of privileges to other entities, respectively.

As shown in Figure 5.4, GSI may be thought of as being composed of four distinct functions: message protection, authentication, delegation, and authorization.

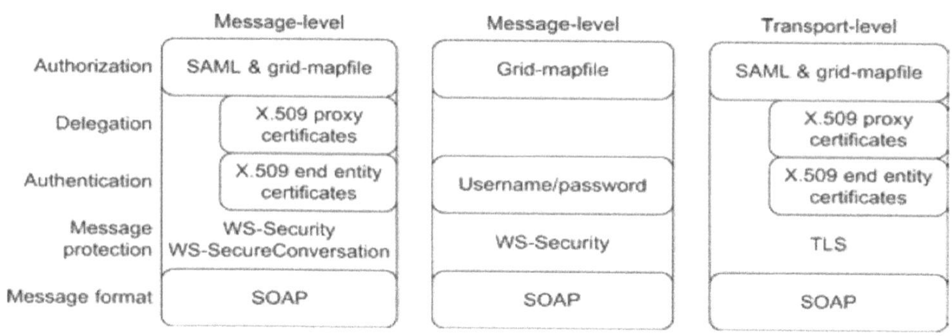

Figure 5.4 GSI functional layers at the message and transport levels

➔ **TLS (transport-level security) or WS-Security and WS-Secure Conversation (message-level)** are used as message protection mechanisms in combination with SOAP.

➔ **X.509 End Entity Certificate**s or Username and Password are used as authentication credentials. X.509 Proxy Certificates and WSTrust are used for delegation.

- ➔ An Authorization Framework allows for a variety of authorization schemes, including a "grid-mapfile" ACL, an ACL defined by a service, a custom authorization handler, and access to an authorization service via the SAML protocol.
- ➔ In addition, associated security tools provide for the storage of X.509 credentials (MyProxy and Delegation services), the mapping between GSI and other authentication mechanisms (e.g., KX509 and PKINIT for Kerberos, MyProxy for one-time passwords), and maintenance of information used for authorization (VOMS, GUMS, PERMIS).
- ➔ The web services portions of GT4 use SOAP as their message protocol for communication. Message protection can be provided either by transport level security, which transports SOAP messages over TLS, or by message-level security, which is signing and/or encrypting portions of the SOAP message using the WS-Security standard.

5.3.2 Transport-Level Security

- Transport-level security entails SOAP messages conveyed over a network connection protected by TLS. TLS provides for both integrity protection and privacy (via encryption).
- Transport-level security is normally used in conjunction with X.509 credentials for authentication, but can also be used without such credentials to provide message protection without authentication, often referred to as "anonymous transport-level security."
- In this mode of operation, authentication may be done by username and password in a SOAP message.

5.3.3 Message-Level Security

- GSI provides message-level security for message protection for SOAP messages by implementing the WS-Security standard and the WS-Secure Conversation specification.
  - ➔ The WSSecurity standard from OASIS defines a framework for applying security to individual SOAP messages
  - ➔ WS-Secure Conversation is a proposed standard from IBM and Microsoft that allows for an initial exchange of messages to establish a security context which can then be used to protect subsequent messages in a manner that requires less computational overhead (i.e., it allows the tradeoff of initial overhead for setting up the session for lower overhead for messages).
- GSI conforms to this standard. GSI uses these mechanisms to provide security on a **per-message basis**, that is, to an individual message without any preexisting context between the sender and receiver (outside of sharing some set of trust roots).
- GSI, as described further in the subsequent section on authentication, allows for both X.509 public key credentials and the combination of username and password for authentication; however, differences still exist.
- With username/password, only the WS-Security standard can be used to allow for authentication; that is, a receiver can verify the identity of the communication initiator.
- GSI allows three additional protection mechanisms.
  - ➔ The first is integrity protection, by which a receiver can verify that messages were not altered in transit from the sender.

- ➔ The second is encryption, by which messages can be protected to provide confidentiality.
- ➔ The third is replay prevention, by which a receiver can verify that it has not received the same message previously.

These protections are provided between WS-Security and WS-Secure Conversation. The former applies the keys associated with the sender and receiver's X.509 credentials. The X.509 credentials are used to establish a session key that is used to provide the message protection.

5.3.4 Authentication and Delegation

GSI has traditionally supported authentication and delegation through the use of X.509 certificates and public keys. As a new feature in GT4, GSI also supports authentication through plain usernames and passwords as a deployment option. We discuss both methods in this section. GSI uses X.509 certificates to identify persistent users and services.

As a central concept in GSI authentication, a certificate includes four primary pieces of information:

1. A subject name, which identifies the person or object that the certificate represents;
2. The public key belonging to the subject;
3. The identity of a CA that has signed the certificate to certify that the public key and the identity both belong to the subject; and
4. The digital signature of the named CA. X.509 provides each entity with a unique identifier (i.e., a distinguished name) and a method to assert that identifier to another party through the use of an asymmetric key pair bound to the identifier by the certificate.

The X.509 certificate used by GSI is conformant to the relevant standards and conventions. Grid deployments around the world have established their own CAs based on third-party software to issue the X.509 certificate for use with GSI and the Globus Toolkit. GSI also supports delegation and single sign-on through the use of standard X.509 proxy certificates. Proxy certificates allow bearers of X.509 to delegate their privileges temporarily to another entity. For the purposes of authentication and authorization, GSI treats certificates and proxy certificates equivalently. Authentication with X.509 credentials can be accomplished either via TLS, in the case of transport-level security, or via signature as specified by WS-Security, in the case of message-level security.

Example: Mutual Authentication between Two Parties

Mutual authentication is processes by which two parties with certificates signed by the CA prove to each other that they are who they say they are based on the certificate and the trust of the CAs that signed each other's certificates.

GSI uses the **Secure Sockets Layer** (SSL) for its mutual authentication protocol, which is described in Figure 5.5. To mutually authenticate, the first person (Alice) establishes a connection to the second person (Bob) to start the authentication process.

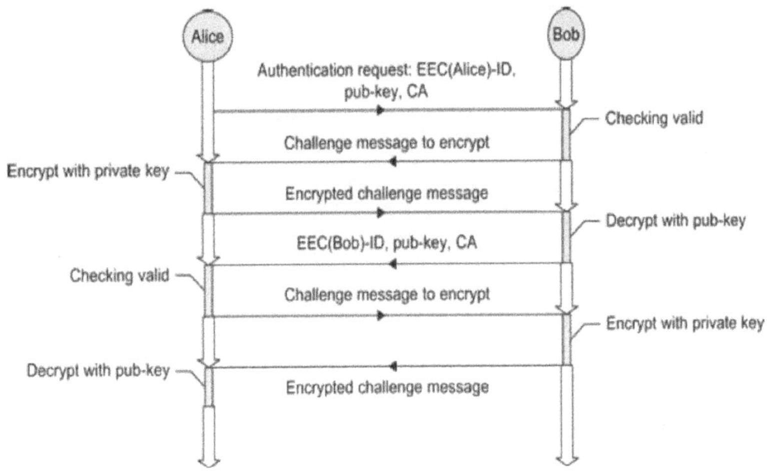

Figure 5.5 Multiple handshaking in a mutual authentication scheme

Alice gives Bob her certificate. The certificate tells Bob who Alice is claiming to be (the identity), what Alice's public key is, and what CA is being used to certify the certificate. Bob will first make sure the certificate is valid by checking the CA's digital signature to make sure the CA actually signed the certificate and the certificate hasn't been tampered with. Once Bob has checked out Alice's certificate, Bob must make sure Alice really is the person identified in the certificate.

Bob generates a random message and sends it to Alice, asking Alice to encrypt it. Alice encrypts the message using her private key, and sends it back to Bob. Bob decrypts the message using Alice's public key. If this results in the original random message, Bob knows Alice is who she says she is. Now that Bob trusts Alice's identity, the same operation must happen in reverse. Bob sends Alice his certificate, and Alice validates the certificate and sends a challenge message to be encrypted. Bob encrypts the message and sends it back to Alice, and Alice decrypts it and compares it with the original. If it matches, Alice knows Bob is who he says he is.

5.3.5 Trust Delegation

To reduce or even avoid the number of times the user must enter his passphrase when several grids are used or have agents (local or remote) requesting services on behalf of a user, GSI provides a delegation capability and a delegation service that provides an interface to allow clients to delegate (and renew) X.509 proxy certificates to a service.

- The interface to this service is based on the WSTrust specification.
- A proxy consists of a new certificate and a private key.
- The key pair that is used for the proxy, that is, the public key embedded in the certificate and the private key, may either be regenerated for each proxy or be obtained by other means.

- The new certificate contains the owner's identity, modified slightly to indicate that it is a proxy.
- The new certificate is signed by the owner, rather than a CA (see Figure 5.6).

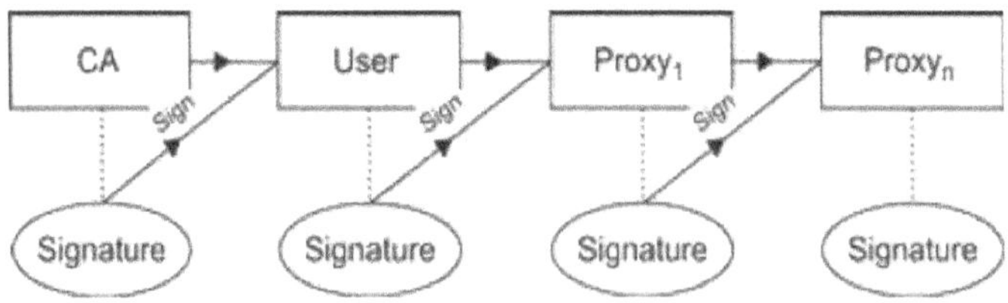

Figure 5.6 A sequence of trust delegations in which new certificates are signed by the owners rather by the CA.

The certificate also includes a time notation after which the **proxy should no longer be accepted by others**.

- **Proxies have limited lifetimes.**

Because the proxy isn't valid for very long, it doesn't have to stay quite as secure as the owner 's private key, and thus it is possible to store the proxy's private key in a local storage system without being encrypted, as long as the permissions on the file prevent anyone else from looking at them easily.

- **Once a proxy is created and stored, the user can use the proxy certificate and private key for mutual authentication without entering a password.**
  - ➔ When proxies are used, the mutual authentication process differs slightly.
  - ➔ The remote party receives not only the proxy's certificate (signed by the owner), but also the owner's certificate.
  - ➔ During mutual authentication, the owner's public key (obtained from her certificate) is used to validate the signature on the proxy certificate.
  - ➔ The CA's public key is then used to validate the signature on the owner's certificate.
  - ➔ This establishes a chain of trust from the CA to the last proxy through the successive owners of resources.

The GSI uses WS-Security with textual usernames and passwords. This mechanism supports more rudimentary web service applications. When using usernames and passwords as opposed to X.509 credentials, the GSI provides authentication, but no advanced security features such as delegation, confidentiality, integrity, and replay prevention. However, one can use usernames and passwords with anonymous transport-level security such as unauthenticated TLS to ensure privacy.

5.4 Cloud Infrastructure security: Network level

When looking at the network level of infrastructure security, it is important to distinguish between public clouds and private clouds.

With private clouds

- There are no new attacks, vulnerabilities, or changes in risk specific to this topology that information security personnel need to consider.
- Although organization's IT architecture may change with the implementation of a private cloud, current network topology will probably not change significantly.
- If you have a private extranet in place (e.g., for premium customers or strategic partners), for practical purposes you probably have the network topology for a private cloud in place already.
- The security considerations you have today apply to a private cloud infrastructure, too.
- And the security tools you have in place (or should have in place) are also necessary for a private cloud and operate in the same way.
- Figure 5.7 shows the topological similarities between a secure extranet and a private cloud.

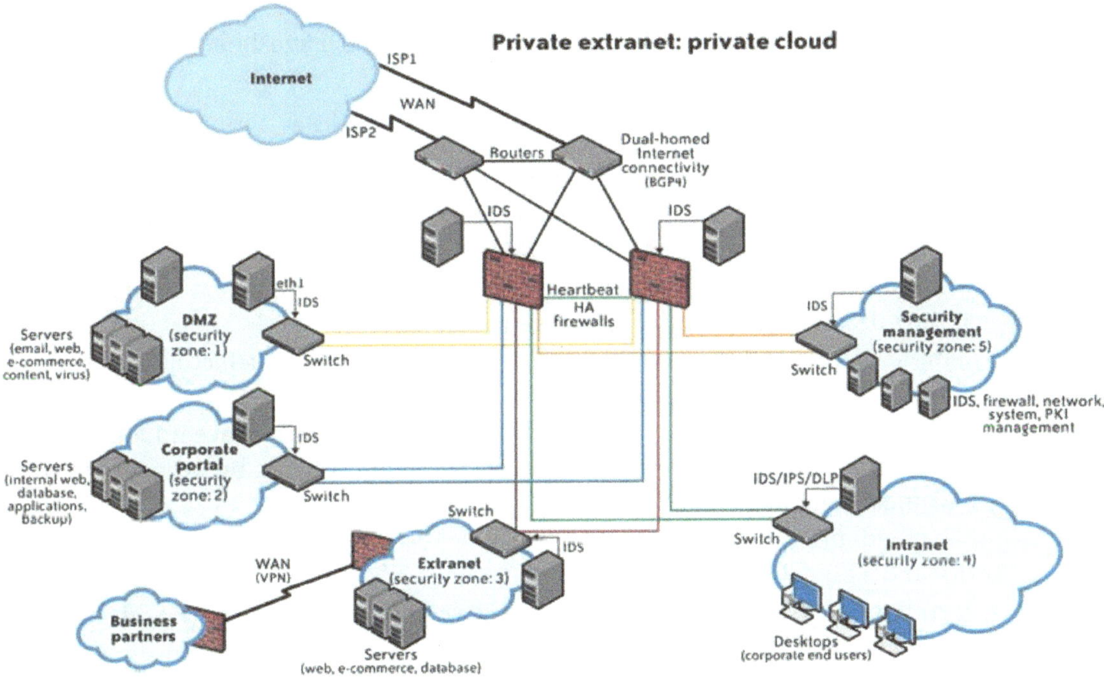

Figure 5.7 Generic network topology for private cloud computing

With public cloud services

- Changing security requirements will require changes to your network topology.
- You must address how your existing network topology interacts with your cloud provider's network topology.
- There are four significant risk factors in this use case:
  - ➔ Ensuring the confidentiality and integrity of your organization's data-in-transit to and from your public cloud provider.
  - ➔ Ensuring proper access control (authentication, authorization, and auditing) to whatever resources you are using at your public cloud provider.

- Ensuring the availability of the Internet-facing resources in a public cloud that are being used by your organization, or have been assigned to your organization by your public cloud providers.
- Replacing the established model of network zones and tiers with domains.

5.4.1 Ensuring data confidentiality and integrity

Some resources and data previously confined to a private network are now exposed to the Internet, and to a shared public network belonging to a third-party cloud provider.

- An example of problems associated with this first risk factor is an Amazon Web Services (AWS) security vulnerability reported in December 2008.* In a blog post, the author detailed a flaw in the digital signature algorithm used when "... making Query (aka REST) requests to Amazon SimpleDB, to Amazon Elastic Compute Cloud (EC2), or to Amazon Simple Queue Service (SQS) over HTTP." Although use of HTTPS (instead of HTTP) would have mitigated the integrity risk, users not using HTTPS (but using HTTP) did face an increased risk that their data could have been altered in transit without their knowledge.

5.4.2 Ensuring proper access control

Since some subset of these resources (or maybe even all of them) is now exposed to the Internet, an organization using a public cloud faces a significant increase in risk to its data. The ability to audit the operations of cloud provider's network is probably non-existent. You will have decreased access to relevant network-level logs and data, and a limited ability to thoroughly conduct investigations and gather forensic data.

- An example of the problems associated with this second risk factor is the issue of reused (reassigned) IP addresses.
- Cloud providers do not sufficiently "age" IP addresses when they are no longer needed for one customer. Addresses are usually reassigned and reused by other customers as they become available.
- From a cloud provider's perspective this makes sense. IP addresses are a finite quantity and a billable asset.
- However, from a customer's security perspective, the persistence of IP addresses those are no longer in use can present a problem.
- A customer can't assume that network access to its resources is terminated upon release of its IP address.
- There is necessarily a lag time between the change of an IP address in DNS and the clearing of that address in DNS caches.
- There is a similar lag time between when physical (i.e., MAC) addresses are changed in ARP tables and when old ARP addresses are cleared from cache; an old address persists in ARP caches until they are cleared.
- This means that even though addresses might have been changed, the (now) old addresses are still available in cache, and therefore they still allow users to reach these supposedly non-existent resources.

- Recently, there were many reports of problems with **"non-aged" IP addresses** at one of the largest cloud providers; this was likely an impetus for an AWS announcement of the Amazon Elastic IP capabilities in March 2008.

However, the issue of "non-aged" IP addresses and unauthorized network access to resources does not apply only to routable IP addresses (i.e., resources intended to be reachable directly from the Internet).

The issue also applies to **cloud providers' internal networks** for customer use and the assignment of non-routable IP addresses.

- Although your resources may not be directly reachable from the Internet, for management purposes your resources must be accessible within the cloud provider's network via private addressing.
- Other customers of your cloud provider may not be well intentioned and might be able to reach your resources internally via the cloud provider's networks.

Some products emerging onto the market will help alleviate the problem of IP address reuse, but unless cloud providers offer these products as managed services, customers are paying for yet another third-party product to solve a problem that their cloud provider's practices created for them.

5.4.3 Ensuring the availability of the Internet-facing resources

Reliance on network security has increased because an increased amount of data or an increased number of organizational personnel now depend on externally hosted devices to ensure the availability of cloud-provided resources. **Misconfiguration** occurs and can block access to data.

Three Risks:

[1]. Prefix Hijacking :

Border Gateway Protocol **prefix hijacking** the falsification of Network Layer Reachability Information provides a good example of this third risk factor. Prefix hijacking involves announcing an autonomous system address space that belongs to someone else without her permission. Such announcements often occur because of a configuration mistake, but that misconfiguration may still affect the availability of your cloud-based resources. The best known example of such a misconfiguration mistake occurred in February 2008 when Pakistan Telecom made an error by announcing a dummy route for YouTube to its own telecommunications partner, PCCW, based in Hong Kong. The intent was to block YouTube within Pakistan because of some supposedly blasphemous videos hosted on the site. The result was that YouTube was globally unavailable for two hours.

[2]. DNS attacks:

Another example of problems associated with this third risk factor. In fact, there are several forms of DNS attacks to worry about with regard to cloud computing. Although DNS attacks are not new and are not directly related to the use of cloud computing, the issue with DNS and cloud computing is an increase in an organization's risk at the network level because of increased external DNS querying (reducing the effectiveness of "split horizon" DNS configurations) along with some increased number of organizational personnel being more dependent on network security to ensure the availability of cloud-provided resources being used.

- **"Kaminsky Bug"**, "DNS Insufficient Socket Entropy Vulnerability" garnered most of the network security attention in 2008, other DNS problems impact cloud computing as well. Not only are there vulnerabilities in the DNS protocol and in implementations of DNS, but also there are fairly widespread DNS
- **Cache poisoning attacks** whereby a DNS server is tricked into accepting incorrect information. Although many people thought DNS cache poisoning attacks had been quashed several years ago, that is not true, and these attacks are still very much a problem—especially in the context of cloud computing.
- Variants of this basic cache poisoning attack include redirecting the target domain's name server (NS), redirecting the NS record to another target domain, and responding before the real NS (called DNS forgery).

[3]. Denial of Service (DoS) and distributed denial of service (DDoS) attacks. DoS/DDoS attacks are not new and are not directly related to the use of cloud computing, the issue with these attacks and cloud computing is an increase in an organization's risk at the network level because of some increased use of resources external to your organization's network. For example, there continue to be rumors of continued DDoS attacks on AWS, making the services unavailable for hours at a time to AWS users.

However, **when using IaaS**, the risk of a DDoS attack is not **only external** (i.e., Internet-facing). There is also the risk of an **internal DDoS attack** through the portion of the IaaS provider's network used by customers (separate from the IaaS provider's corporate network). That internal (non-routable) network is a shared resource, used by customers for access to their non-public instances (e.g., Amazon Machine Images or AMIs) as well as by the provider for management of its network and resources (such as physical servers.

The only **preventive controls** other customers would have would be how hardened their instances (e.g., AMIs) are, and whether they are taking advantage of a provider's capabilities to firewall off groups of instances (e.g., AWS).

5.4.4 Replacing the established model of network zones and tiers with domains

The established isolation model of network zones and tiers no longer exists in the public IaaS and PaaS clouds. For years, network security has relied on zones, such as **intranet versus extranet and development versus production, to segregate network traffic** for improved security. This model was based on **exclusion** only individuals and systems in specific roles have access to specific zones. Similarly, systems within a specific tier often have only specific access within or across a specific tier.

- For example, systems within a presentation tier are not allowed to communicate directly with systems in the database tier, but can communicate only with an authorized system within the application zone. SaaS clouds built on public IaaS or PaaS clouds have similar characteristics. However, a public SaaS built on a private IaaS (e.g., Salesforce.com) may follow the traditional isolation model, but that topology information is not typically shared with customers.

The traditional model of network zones and tiers has been replaced in public cloud computing with **"security groups," "security domains "or" virtual data centers"** that have logical separation between tiers but are less precise and afford less protection than the formerly established model.

- For example, the security groups feature in AWS allows your virtual machines (VMs) to access each other using a virtual firewall that has the ability to filter traffic based on IP address (a specific address or a subnet), packet types (TCP, UDP, or ICMP), and ports (or a range of ports). Domain names are used in various networking contexts and application-specific naming and addressing purposes, based on DNS. For example, Google's App Engine provides a logical grouping of applications based on domain names such as mytestapp.test.mydomain.com and myprodapp.prod.mydomain.com.

In the established model of network zones and tiers, not only were **development systems logically separated from production systems at the network level**, but these two groups of systems were also physically separated at the host level (i.e., they ran on physically separated servers in logically separated network zones). With cloud computing, however, this separation no longer exists. The cloud computing model of separation by domains provides logical separation for addressing purposes only. There is no longer any "required" physical separation, as a test domain and a production domain may very well be on the same physical server.

Furthermore, the former logical network separation no longer exists; logical separation now is at the host level with both domains running on the same physical server and being separated only logically by VM monitors (hypervisors).

5.4.5 Network Level Migration

What can you do to mitigate these increased risk factors?

- First, note that network-level risks exist regardless of **what aspects of "cloud computing" services are being used** (e.g., software-as-a-service, platform-as-a-service, or infrastructure-as-a-service).
  - ➔ The primary determination of risk level is therefore not which *aaS is being used, but rather whether your organization intends to use or is using a public, private, or hybrid cloud.
  - ➔ Although some IaaS clouds offer virtual network zoning, they may not match an internal private cloud environment that performs stateful inspection and other network security measures.
- If your organization is large enough to **afford the resources of a private cloud**, your risks will decrease assuming you have a true private cloud that is internal to your network.
  - ➔ In some cases, a private cloud located at a cloud provider's facility can help meet your security requirements but will depend on the provider capabilities and maturity.
- You can **reduce your confidentiality risks by using encryption**; specifically by using validated implementations of cryptography for data-in-transit.
  - ➔ Secure digital signatures make it much more difficult, if not impossible, for someone to tamper with your data, and this ensures data integrity.

- **Availability problems** at the network level are far more difficult to mitigate with cloud computing unless your organization is using a private cloud that is internal to your network topology.
  - ➔ Even if your private cloud is a private (i.e., non-shared) external network at a cloud provider's facility, you will face increased risk at the network level. A public cloud faces even greater risk.
- Even large enterprises with significant resources face considerable **challenges at the network level of infrastructure security**.
  - ➔ Are the risks associated with cloud computing actually higher than the risks enterprises are facing today? Consider existing private and public extranets, and take into account partner connections when making such a comparison.
  - ➔ For large enterprises without significant resources, or for small to medium-size businesses (SMBs), is the risk of using public clouds (assuming that such enterprises lack the resources necessary for private clouds) really higher than the risks inherent in their current infrastructures? In many cases, the answer is probably no—there is not a higher level of risk.

Threat outlook	**Low (with the exception of DoS attacks)**
Preventive controls	Network access control supplied by provider (e.g., firewall), encryption of data in transit (e.g., SSL, IPSec)
Detective controls	Provider-managed aggregation of security event logs (security incident and event management, or SIEM), network-based intrusion detection system/intrusion prevention system (IDS/IPS)

Table 5.1 Security Control at the network level

5.5 Infrastructure security: Host level

When reviewing host security and assessing risks consider the context of cloud services delivery models (SaaS, PaaS, and IaaS) and deployment models (public, private, and hybrid).

- Some **virtualization security threats** such as VM escape, system configuration drift, and insider threats by way of weak access control to the hypervisor carry into the public cloud computing environment.
- The dynamic nature (elasticity) of cloud computing bring **new operational challenges** from a security management perspective.
  - ➔ The operational model motivates rapid provisioning and fleeting instances of VMs. Managing vulnerabilities and patches is therefore much harder than just running a scan, as the rate of change is much higher than in a traditional data center.
- The fact that the clouds harness the power of thousands of compute nodes, combined with the homogeneity of the operating system employed by hosts, means the threats can be amplified quickly and easily calls it the **"velocity of attack"** factor in the cloud.

5.5.1 SaaS and PaaS Host Security

In general, CSPs **do not publicly share information** related to their host platforms, host operating systems, and the processes that are in place to secure the hosts, since hackers can exploit that information when they are trying to intrude into the cloud service.

- Hence, in the context of SaaS (e.g., Salesforce.com, Workday.com) or PaaS (e.g., Google App Engine, Salesforce.com's Force.com) cloud services, host security is solid to customers and the responsibility of securing the hosts is relegated to the CSP.

To get assurance from the CSP on the security hygiene of its hosts, you should ask the vendor to share information under a **nondisclosure agreement** (NDA) or simply demand that the CSP share the information via a controls assessment framework such as SysTrust or ISO 27002.

- From a controls assurance perspective, the CSP has to ensure that appropriate preventive and detective controls are in place and will have to ensure the same via a third-party assessment or ISO 27002 type assessment framework.

Both the PaaS and SaaS platforms **abstract and hide the host operating system from end users** with a host abstraction layer.

- One key difference between PaaS and SaaS is the accessibility of the abstraction layer that hides the operating system services the applications consume.
- In the case of SaaS, the abstraction layer is not visible to users and is available only to the developers and the CSP's operations staff,
- where PaaS users are given indirect access to the host abstraction layer in the form of a PaaS application programming interface (API) that in turn interacts with the host abstraction layer.

  Host security responsibilities in SaaS and PaaS services are transferred to the CSP.
- The fact that you do not have to worry about protecting hosts from host-based security threats is a major benefit from a security management and cost standpoint.
- However, as a customer, you still own the risk of managing information hosted in the cloud services.
- It's your responsibility to get the appropriate level of assurance regarding how the CSP manages host security hygiene.

5.5.2 IaaS Host Security

Unlike PaaS and SaaS, IaaS customers are primarily responsible for securing the hosts provisioned in the cloud. Given that almost all IaaS services available today employ virtualization at the host layer, host security in IaaS should be categorized as follows:

Virtualization software security

The software layer that sits on top of bare metal and provides customers the ability to create and destroy virtual instances.

Virtualization at the host level can be accomplished using any of the virtualization models, including

- OS-level virtualization (Solaris containers, BSD jails, Linux-VServer)
- Paravirtualization (a combination of the hardware version and versions of Xen and VMware)

- Hardware-based virtualization (Xen, VMware, Microsoft Hyper-V).

It is important to secure this layer of software that sits between the hardware and the virtual servers.

In a public IaaS service, customers do not have access to this software layer; it is managed by the CSP only.

Customer guest OS or virtual server security

The virtual instance of an operating system that is provisioned on top of the virtualization layer and is visible to customers from the Internet; e.g., various flavors of Linux, Microsoft, and Solaris. Customers have full access to virtual servers.

5.5.2.1 Virtualization software security

Since the CSP manages the virtualization software that sits on top of the hardware, customers will have neither visibility nor access to this software.

- Hardware or OS virtualization enables the sharing of hardware resources across multiple guest VMs without interfering with each other so that you can safely run several operating systems and applications at the same time on a single computer.
- For the purpose of simplicity, we made an assumption that IaaS services are using "bare metal hypervisor" technologies.

Given that **hypervisor virtualization** is the essential ingredient that guarantees compartmentalization and isolation of customer VMs from each other in a multitenant environment, it is very important to protect the hypervisors from unauthorized users.

- Since virtualization is very critical to the IaaS cloud architecture, any attack that could compromise the integrity of the compartments will be catastrophic to the entire customer base on that cloud.
- By exploiting a zero-day vulnerability in HyperVM, a virtualization application made by a company called Lxlabs, hackers destroyed 100,000 websites hosted by Vaserv.com.

The **zero-day vulnerability** gave the attackers the ability to execute sensitive Unix commands on the system, including rm -rf, which forces a recursive delete of all files.

- Evidently, just days before the intrusion, an anonymous user posted on a hacker website called milw0rm a long list of yet-unpatched vulnerabilities in Kloxo, a hosting control panel that integrates into HyperVM.

CSPs should institute the necessary security controls, including restricting physical and logical access to hypervisor and other forms of employed virtualization layers. IaaS customers should understand the technology and security process controls instituted by the CSP to protect the hypervisor. This will help you to understand the compliance and gaps with reference to your host security standard, policies, and regulatory compliances. However, in general, CSPs lack transparency in this area and you may have no option but to take a leap of faith and trust CSPs to provide an "isolated and secured virtualized guest OS."

5.5.2.1.1 Threads to the hypervisor

The **integrity and availability of the hypervisor** are of utmost importance and are key to guaranteeing the integrity and availability of a public cloud built on a virtualized environment.

A **vulnerable hypervisor** could expose all user domains to malicious insiders. Furthermore, hypervisors are potentially susceptible to subversion attacks.

- To illustrate the vulnerability of the virtualization layer, some members of the security research community demonstrated a **"Blue Pill" attack on a hypervisor**. During Black Hat 2008 and Black Hat DC 2009,Joanna Rutkowska, Alexander Tereshkin, and Rafal Wojtczuk from Invisible Things Lab demonstrated a number of ways to compromise Xen's virtualization.

Since virtualization layers within public clouds for the most part are **proprietary and closed source**, the source code of software used by CSPs is not available for scrutiny by the security research community.

5.5.2.1.2 Virtual server security

- Customers of IaaS have full access to the virtualized guest VMs that are hosted and isolated from each other by hypervisor technology.
    - ➔ Hence customers are responsible for securing and ongoing security management of the guest VM.
- A public IaaS, offers a web services API to perform management functions such as provisioning, decommissioning, and replication of virtual servers on the IaaS platform.
- The dynamic life cycle of virtual servers can result in complexity if the process to manage the virtual servers is not automated with proper procedures.
    - ➔ From an attack surface perspective, the virtual server (Windows, Solaris, or Linux) may be accessible to anyone on the Internet, so sufficient network access mitigation steps should be taken to restrict access to virtual instances.
- The cloud management API adds another layer of attack surface and must be included in the scope of securing virtual servers in the public cloud.
- Some of the new host security threats in the public IaaS include:
    - ➔ Stealing keys used to access and manage hosts (e.g., SSH private keys)
    - ➔ Attacking un patched, vulnerable services listening on standard ports (e.g., FTP, NetBIOS,SSH)
    - ➔ Hijacking accounts that are not properly secured (i.e., weak or no passwords for standard accounts)
    - ➔ Attacking systems that are not properly secured by host firewalls
    - ➔ Deploying Trojans embedded in the software component in the VM or within the VM image (the OS) itself

5.5.2.2 Securing Virtual server

Securing the virtual server in the cloud requires strong operational security procedures coupled with automation of procedures. Here are some recommendations:

- **Use a secure-by-default configuration**. Harden your image and use a standard hardened image for instantiating VMs (the guest OS) in a public cloud. A best practice for cloud based applications is to build custom VM images that have only the capabilities and services necessary to support the application stack. Limiting the capabilities of the underlying application stack not only limits the host's overall attack surface, but also greatly reduces the number of patches needed to keep that application stack secure.

- **Track the inventory of VM images and OS versions** that are prepared for cloud hosting. The IaaS provider provides some of these VM images. When a virtual image from the IaaS provider is used it should undergo the same level of security verification and hardening for hosts within the enterprise. The best alternative is to provide your own image that conforms to the same security standards as internal trusted hosts.
- **Protect the integrity** of the hardened image from unauthorized access.
- **Safeguard the private keys** required to access hosts in the public cloud.
- **Isolate the decryption keys** from the cloud where the data is hosted unless they are necessary for decryption, and then only for the duration of an actual decryption activity. If your application requires a key to encrypt and decrypt for continuous data processing, it may not be possible to protect the key since it will be collocated with the application.
- **Include no authentication credentials** in your virtualized images except for a key to decrypt the file 6system key.
- **Do not allow password-based** authentication for shell access.
- **Require passwords** for sudo or role-based access (e.g., Solaris, SELinux).
- **Run a host firewall** and open only the minimum ports necessary to support the services on an instance.
- **Run only the required services** and turn off the unused services (e.g., turn off FTP, print services, network file services, and database services if they are not required).
- **Install host-based IDS** such as OSSEC or Samhain.
- **Enable system auditing and event logging**, and log the security events to a dedicated log server. Isolate the log server with higher security protection, including accessing controls.
- If you **suspect a compromise**, shut down the instance, snapshot your block volumes, and back up the root file system. You can perform forensics on an uncompromised system later.
- **Institute a process for patching the images** in the cloud—both offline and instantiated images.
- **Periodically review** logs for suspicious activities.
- Table 5.2 lists security controls at the host level.

Table 5.2 Security controls at the host levels

Threat outlook	High
Preventive controls	Host firewall, access control, patching, hardening of system, strong authentication
Detective controls	Security event logs, host-based IDS/IPS

5.6 Infrastructure security: Application level

- Designing and implementing applications targeted for deployment on a cloud platform will require that existing application security programs reevaluate current practices and standards.
- The application security ranges from standalone single-user applications to sophisticated multiuser e-commerce applications used by millions of users.
- Web applications such as content management systems (CMSs), wikis, portals, bulletin boards, and discussion forums are used by small and large organizations.
- A large number of organizations also develop and maintain custom-built web applications for their businesses using various web frameworks.
- Advances in cross-site scripting (XSS) attacks have demonstrated that criminals looking for financial gain can exploit vulnerabilities resulting from web programming errors as new ways to penetrate important organizations.

Since the browser has emerged as the end user client for accessing in-cloud applications, it is important for application security programs to include browser security into the scope of application security. Together they determine the strength of end-to-end cloud security that helps protect the confidentiality, integrity, and availability of the information processed by cloud services.

5.6.1 Application –level security threads

Web application vulnerabilities in open source as well as custom-built applications accounted for almost half the total number of vulnerabilities discovered between November 2006 and October 2007.

- The existing threats exploit well-known application vulnerabilities including
  - ➔ cross-site scripting (XSS),
  - ➔ SQL injection,
  - ➔ malicious file execution, and
  - ➔ Other vulnerabilities resulting from programming errors and design flaws.
- Armed with knowledge and tools, hackers are constantly scanning web applications for application vulnerabilities.
- They are then exploiting the vulnerabilities they discover for various illegal activities including
  - ➔ financial fraud,
  - ➔ intellectual property theft,
  - ➔ converting trusted websites into malicious servers serving client-side exploits,
  - ➔ and phishing scams.

- All web frameworks and all types of web applications are at risk of web application security defects, ranging from insufficient validation to application logic errors.
- It has been a common practice to **use a combination of perimeter security controls and network- and host-based access controls** to protect web applications deployed in a tightly controlled environment, including corporate intranets and private clouds, from external hackers.
- Web applications built and deployed in a public cloud platform will be subjected to a high threat level, attacked, and potentially exploited by hackers to support fraudulent and illegal activities.

In that threat model, web applications deployed in a public cloud (the SPI model) must be designed for an Internet threat model, and security must be embedded into the Software Development Life Cycle (SDLC); see Figure 5.8.

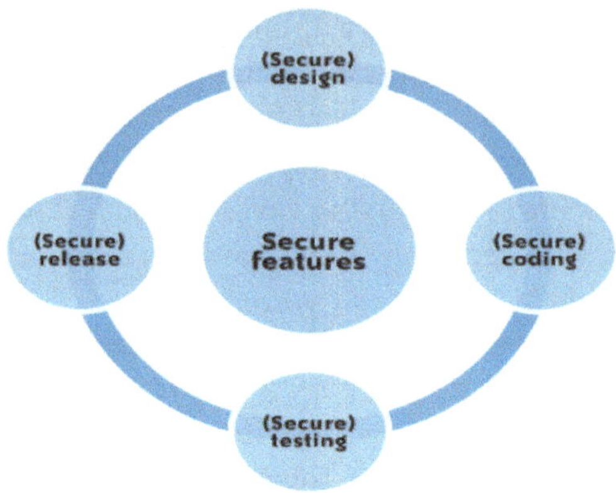

Figure 5.8 The SDLC

5.6.2 DoS and EDoS

- DoS and DDoS attacks that can potentially disrupt cloud services for an extended time.
- These attacks typically originate from compromised computer systems attached to the Internet
  - ➔ routinely,
  - ➔ hackers hijack and control computers infected by way of viruses/worms/malware and,
  - ➔ in some cases, powerful unprotected servers
- Application-level DoS attacks could manifest themselves as high-volume web page reloads, XML web services requests (over HTTP or HTTPS), or protocol-specific requests supported by a cloud service.
- Since these malicious requests blend with the legitimate traffic, it is extremely difficult to selectively filter the malicious traffic without impacting the service as a whole.
  - ➔ For example, a DDoS attack on Twitter on August 6, 2009, brought the service down for several hours (see Figure 5.9).

Apart from disrupting cloud services, resulting in poor user experience and service-level impacts, DoS attacks can quickly drain your company's cloud services budget.

- DoS attacks on pay-as-you-go cloud applications will result in a dramatic increase in your cloud utility bill: you'll see increased use of network bandwidth, CPU, and storage consumption. This type of attack is also being characterized as **economic denial of sustainability (EDoS).**
- The low barriers for small and medium-size enterprises to adopt cloud computing for legitimate use are also leveling the field for hackers.
- Using hijacked or exploited cloud accounts, hackers will be able to link together computing resources to achieve massive amounts of computing without any of the capital infrastructure costs.

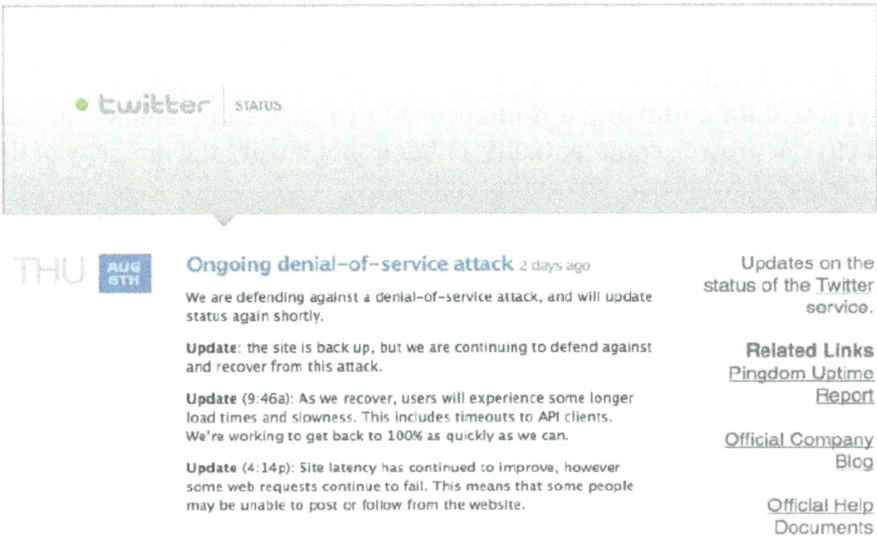

Figure 5.9 DoS attack on Twitter

5.6.3 End user security

- A customer of a cloud service, are responsible for end user security tasks like security procedures to protect your Internet-connected PC—and for practicing "safe surfing."
- Protection measures include use of security software, such as anti-malware, antivirus, personal firewalls, security patches, and IPS-type software on your Internet-connected computer.
- The new mantra of "the browser is your operating system" appropriately conveys the message that browsers have become the ubiquitous "operating systems" for consuming cloud services.
- All Internet browsers routinely suffer from software vulnerabilities that make them vulnerable to end user security attacks.
    - Hence, our recommendation is that cloud customers take appropriate steps to protect browsers from attacks.
    - To achieve end-to-end security in a cloud, it is essential for customers to maintain good browser hygiene. The means keeping the browser (e.g., Internet Explorer, Firefox, Safari) patched and updated to mitigate threats related to browser vulnerabilities.

- Currently, although browser security add-ons are not commercially available, users are encouraged to frequently check their browser vendor's website for security updates, use the auto-update feature, and install patches on a timely basis to maintain end user security.

5.7 Aspects of data security

With regard to data-in-transit, the primary risk is in not using a vetted encryption algorithm.

- It is also important to ensure that a protocol provides **confidentiality as well as integrity** (e.g., FTP over SSL [FTPS], Hypertext Transfer Protocol Secure [HTTPS], and Secure Copy Program [SCP])—particularly if the protocol is used for transferring data across the Internet.
- **Encrypting data and using a non-secured protocol** (e.g., "vanilla" or "straight" FTP or HTTP) can provide confidentiality, but does not ensure the integrity of the data (e.g., with the use of symmetric streaming ciphers).

Using encryption to protect data-at-rest might seem obvious, the reality is not that simple.

- If you are using an IaaS cloud service (public or private) for simple storage encrypting data-at-rest is possible and is strongly suggested.
- However, encrypting data-at-rest that a PaaS or SaaS cloud-based application is using (e.g., Google Apps, Salesforce.com) as a compensating control is not always feasible.
- Data-at-rest used by a cloud-based application is generally not encrypted, because encryption
  would prevent indexing or searching of that data.

Data, when processed by a cloud-based application or stored for use by a cloud-based application, is commingled with other users' data.

- Although applications are often designed with features such as data tagging to prevent unauthorized access to commingled data, unauthorized access is still possible through some exploit of application vulnerability.
- Although some cloud providers have their applications reviewed by third parties or verified with third-party application security tools, data is not on a platform dedicated solely to one organization.

An organization's data-in-transit might be encrypted during transfer to and from a cloud provider, and its data-at-rest might be encrypted if using simple storage.

- Organization's data is definitely not encrypted if it is processed in the cloud (public or private).
- For any application to process data, that data must be unencrypted.
- Therefore, unless the data is in the cloud for only simple storage, the data will be unencrypted during at least part of its life cycle in the cloud—processing at a minimum.
    - In June 2009, IBM announced that one of its researchers, working with a graduate student from Stanford University, had developed a fully homomorphic encryption scheme which allows data to be processed without being decrypted. This is a huge advance in cryptography, and it will have a

significant positive impact on cloud computing as soon as it moves into Whether the data an organization has put into the cloud is encrypted or not, it is useful and might be required to know exactly where and when the data was specifically located within the cloud.

For example, the data might have been transferred to a cloud provider, such as Amazon Web Services (AWS), on date x1 at time y1 and stored in a bucket on Amazon's S3 in example1.s3.amazonaws.com, then processed on date x2 at time y2 on an instance being used by an organization on Amazon's Elastic Compute Cloud (EC2) in ec2-67-202-51-223.compute-1.amazonaws.com, then restored in another bucket, example2.s3.amazonaws.com, before being brought back into the organization for storage in an internal data warehouse belonging to the marketing operations group on date x3 at time y3.

Data lineage

Following the path of data mapping application data flows or data path visualization is known as **data lineage**, and it is important for an auditor's assurance (internal, external, and regulatory). However, providing data lineage to auditors or management is **time-consuming**, even when the environment is completely under an organization's control. Trying to provide accurate reporting on data lineage for a public cloud service is really not possible. In the preceding example, on what physical system is that bucket on example1.s3.amazonaws.com, and specifically where is (or was) that system located? What was the state of that physical system then, and how would a customer or auditor verify that information?

Proving data provenance

Even if data lineage can be established in a public cloud, for some customers there is an even more challenging requirement and problem: proving data provenance—not just proving the integrity of the data, but the more specific provenance of the data.

There is an important difference between the two terms.

- **Integrity** of data refers to data that has not been changed in an unauthorized manner or by an unauthorized person.
- **Provenance** means not only that the data has integrity, but also that it is computationally accurate; that is, the data was accurately calculated.

For example, consider the following financial equation:

SUM ((((2*3)*4)/6)-2) = \$2.00

With that equation, the expected answer is \$2.00. If the answer were different, there would be an integrity problem. Of course, the assumption is that the \$2.00 is in U.S. dollars, but the assumption could be incorrect if a different dollar is used with the following associated assumptions:

- The equation is specific to the Australian, Bahamian, Barbadian, Belize, Bermudian, Brunei, Canadian, Cayman Islands, Cook Islands, East Caribbean, Fijian, Guyanese, Hong Kong, Jamaican, Kiribati, Liberian, Namibian, New Zealand, Samoan, Singapore, Solomon Islands, Surinamese, New Taiwan, Trinidad and Tobago, Tuvaluan, or Zimbabwean dollar.
- The dollar is meant to be converted from another country's dollars into U.S. dollars.

- The correct exchange rate is used and the conversion is calculated correctly and can be proven.

In this example, if the equation satisfies those assumptions, the equation has integrity but not provenance.

Data remanence

A final aspect of data security is data remanence. "Data remanence is the residual representation of data that has been in some way nominally erased or removed. This residue may be due to data being left intact by a nominal delete operation, or through physical properties of the storage medium. Data remanence may make inadvertent disclosure of sensitive information possible, should the storage media be released into an uncontrolled environment (e.g., thrown in the trash, or given to a third party).

- The risk posed by data remanence in cloud services is that an organization's data can be inadvertently exposed to an unauthorized party—regardless of which cloud service you are using (SaaS, PaaS, or IaaS).

In spite of the increased importance of data security, the attention that cloud service providers (CSPs) pay to data remanence is strikingly low. Many do not even mention data remanence in their services.

The two approved methods of data (destruction) security, but does not provide any specific requirements for how these two methods are to be achieved, nor does it provide any standards for how these methods are to be accomplished.

Data remanence is limited to three paragraphs:

"Clearing and Sanitization"- Instructions on clearing, sanitization, and release of information systems (IS) media shall be issued by the accrediting Cognizant Security Agency (CSA).

- **"Clearing"**- Clearing is the process of eradicating the data on media before reusing the media in an environment that provides an acceptable level of protection for the data that was on the media before clearing. All internal memory, buffer, or other reusable memory shall be cleared to effectively deny access to previously stored information.
- **"Sanitization"-** Sanitization is the process of removing the data from media before reusing the media in an environment that does not provide an acceptable level of protection for the data that was on the media before sanitizing. IS resources shall be sanitized before they are released from classified information controls or released for use at a lower classification level.

5.7.1 Data security migration

- If prospective customers of cloud computing services expect that data security will serve as compensating controls for possibly weakened infrastructure security,
    - since part of a customer's infrastructure security moves beyond its control and
    - a provider's infrastructure security may (for many enterprises) or may not (for small to medium-size businesses, or SMBs) be less robust than expectations, you will be disappointed.

- Although data-in-transit can and should be encrypted, any use of that data in the cloud, beyond simple storage, requires that it be decrypted.

Therefore, it is almost certain that in the cloud, data will be unencrypted. And if you are using a PaaS-based application or SaaS, customer-unencrypted data will also almost certainly be hosted in a multitenancy environment (in public clouds). Add to that exposure the difficulties in determining the data's lineage, data provenance where necessary and even many providers' failure to adequately address such a basic security concern as data remanence, and the risks of data security for customers are significantly increased.

The only viable option for mitigation is to **ensure that any sensitive or regulated data is not placed into a public cloud** (or that you encrypt data placed into the cloud for simple storage only). It may be that those economics change and that providers offer their current services, as well as a "regulatory cloud environment" (i.e., an environment where customers are willing to pay more for enhanced security controls to properly handle sensitive and regulated data). Currently, the only viable option for mitigation is to ensure that any sensitive or regulated data is not put into a public cloud.

5.7.2 Provider data and its security

In addition to the security of your own customer data, customers should also be concerned about what data the provider collects and how the CSP protects

- customer data,
- what metadata does the provider have about your data,
- how is it secured, and
- What access do you, the customer, have to that metadata?
- As your volume of data with a particular provider increases, so does the value of that metadata.

Additionally, provider collects and must protect a huge amount of security-related data.

- At the **network level**, your provider should be collecting, monitoring, and protecting firewall, intrusion prevention system (IPS), security incident and event management (SIEM), and router flow data.
- At the **host level** your provider should be collecting system log files, and at the application level SaaS providers should be collecting application log data, including authentication and authorization information.

What data your CSP collects and how it monitors and protects that data is important to the provider for its own audit purposes. Additionally, this information is important to both providers and customers in case it is needed for incident response and any digital forensics required for incident analysis.

5.7.3 Storage

For data stored in the cloud (i.e., storage-as-a-service), we are referring to IaaS and not data associated with an application running in the cloud on PaaS or SaaS. The same three information security concerns are associated with this data stored in the as with data stored elsewhere: confidentiality, integrity, and availability.

5.7.4 Confidentiality

When it comes to the confidentiality of data stored in a public cloud have two potential concerns.

- **First, what access control exists to protect the data?** Access control consists of both authentication and authorization. CSPs generally use weak authentication mechanisms (e.g., username + password), and the authorization ("access") controls available to users tend to be quite coarse and not very granular.
- For large organizations, this coarse authorization presents significant security concerns unto itself.
    - ➔ Often, the only authorization levels cloud vendors provide are administrator authorization (i.e., the owner of the account itself) and
    - ➔ user authorization (i.e., all other authorized users)—with no levels in between (e.g., business unit administrators, who are authorized to approve access for their own business unit personnel)
- **Second, how is the data that is stored in the cloud actually protected?** For all practical purposes, protection of data stored in the cloud involves the use of encryption. So, is a customer's data actually encrypted when it is stored in the cloud? And if so, with what encryption algorithm, and with what key strength?
- **If a CSP does encrypt a customer's data, the next consideration concerns what encryption algorithm it uses.**

Although the example in Figure 5.10 is related to email, the same concept (i.e., a single shared, secret key) is used in data storage encryption.

Figure 5.10 symmetric Encryption

Although the example in Figure 5.11 is related to email, the same concept (i.e., a public key and a private key) is not used in data storage encryption.

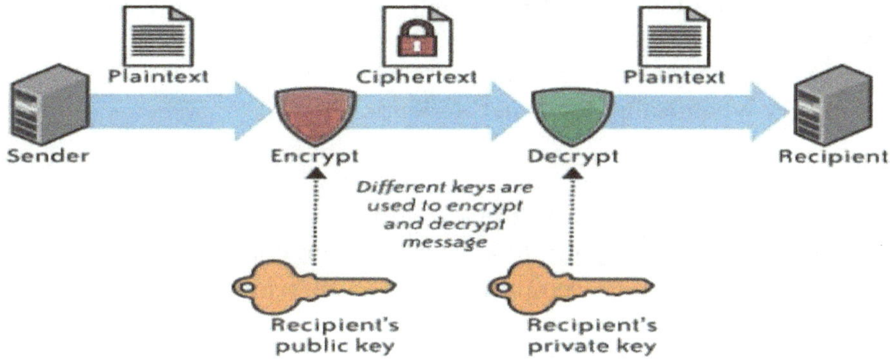

Figure 5.11 Asymmetric Encryption

- The next consideration for you is what key length is used.
- Another confidentiality consideration for encryption is key management. How are the encryption keys that are used going to be managed—and by whom? Are you going to manage your own keys?
  "Recommendation for Key Management":
  - ➔ "Part 1: General"
  - ➔ "Part 2: Best Practices for Key Management Organization"
  - ➔ "Part 3: Application-Specific Key Management Guidance (Draft)"

Key management is complex and difficult for a single customer, it is even more complex and difficult for CSPs to try to properly manage multiple customers' keys.

5.7.5 Integrity

Confidentiality does not imply integrity; data can be encrypted for confidentiality purposes, and yet you might not have a way to verify the integrity of that data. Encryption alone is sufficient for confidentiality, but integrity also requires the use of message authentication codes (MACs).

- ➔ The simplest way to use MACs on encrypted data is to use a block symmetric algorithm in cipher block chaining (CBC) mode, and to include a one-way hash function.
- ➔ that not all providers encrypt customer data, especially for PaaS and SaaS services.

Another aspect of data integrity is important, especially with bulk storage using IaaS. Once a customer has several gigabytes (or more) of its data up in the cloud for storage, how does the customer check on the integrity of the data stored there?

- ➔ There are IaaS transfer costs associated with moving data into and back down from the cloud, as well as network utilization (bandwidth) considerations for the customer's own network.
- ➔ What a customer really wants to do is to validate the integrity of its data while that data remains in the cloud without having to download and reupload that data.

This task is even more difficult because it must be done in the cloud without explicit knowledge of the whole data set. Customers generally do not know on which physical machines their data is stored, or where those systems are located. Additionally, that data set is probably dynamic and changing frequently. Those frequent changes obviate the effectiveness of traditional integrity insurance techniques.

What is needed instead is a proof of retrievability—that is, a mathematical way to verify the integrity of the data as it is dynamically stored in the cloud.

5.7.6 Availability

Assuming that a customer's data has maintained its confidentiality and integrity, you must also be concerned about the availability of your data. There are currently three major threats in this regard none of which are new to computing, but all of which take on increased importance in cloud computing because of increased risk.

- ➔ The first threat to availability is network-based attacks.
- ➔ The second threat to availability is the CSP's own availability.
- ➔ Finally, prospective cloud storage customers must be certain to ascertain just what services their provider is actually offering

All three of these considerations (confidentiality, integrity, and availability) should be encapsulated in a CSP's service-level agreement (SLA) to its customers. However, at this time, CSP SLAs are extremely weak—in fact, for all practical purposes, they are essentially worthless. Even where a CSP appears to have at least a partially sufficient SLA, how that SLA actually gets measured is problematic.

5.8 Introduction to Identity and access management architecture

Identity and Access management (IAM) and support for IAM features that aid in Authentication, Authorization, and Auditing (AAA) of users accessing cloud services. Traditionally, organizations invest in IAM practices to improve operational efficiency and to comply with regulatory, privacy, and data protection requirements:

Improve operational efficiency: Properly architected IAM technology and processes can improve efficiency by automating user on-boarding and other repetitive tasks (e.g., self-service for users requesting password resets that otherwise will require the intervention of system administrators using a help desk ticketing system).

Regulatory compliance management: To protect systems, applications, and information from internal and external threats (e.g., disgruntled employees deleting sensitive files) and to comply with various regulatory, privacy, and data protection requirements (e.g., HIPAA, SOX), organizations implement an "IT general and application-level controls" framework derived from industry standard.

In addition to improving operational efficiencies and effective compliance management, IAM can enable new IT delivery and deployment models. Some of the cloud use cases that require IAM support from the CSP include:

- Employees and on-site contractors of an organization accessing a SaaS service using identity federation.
- IT administrators accessing the CSP management console to provision resources and access for users using a corporate identity.
- Developers creating accounts for partner users in a PaaS platform.

- End users accessing storage service in the cloud (e.g., Amazon S3) and sharing files and objects with users, within and outside a domain using access policy management features.
- An application residing in a cloud service provider (e.g., Amazon EC2) accessing storage from another cloud service (e.g., Mosso)

5.8.1 IAM definitions

The basic concepts and definitions of IAM functions for any service:

Authentication: Authentication is the process of verifying the identity of a user or system. Authentication usually connotes a more robust form of identification. In some use cases, such as service-to-service interaction, authentication involves verifying the network service requesting access to information served by another service (e.g., a travel web service that is connecting to a credit card gateway to verify the credit card on behalf of the user).

Authorization: Authorization is the process of determining the privileges the user or system is entitled to once the identity is established. In the context of digital services, authorization usually follows the authentication step and is used to determine whether the user or service has the necessary privileges to perform certain operations, authorization is the process of enforcing policies.

Auditing: In the context of IAM, auditing entails the process of review and examination of authentication, authorization records, and activities to determine the adequacy of IAM system controls, to verify compliance with established security policies and procedures (e.g., separation of duties), to detect breaches in security services (e.g., privilege escalation), and to recommend any changes that are indicated for countermeasures.

5.8.2 IAM functional architecture and Practices

IAM is not a monolithic solution that can be easily deployed to gain capabilities immediately. It is as much an aspect of architecture (see Figure 5.12) as it is a collection of technology components, processes, and standard practices.

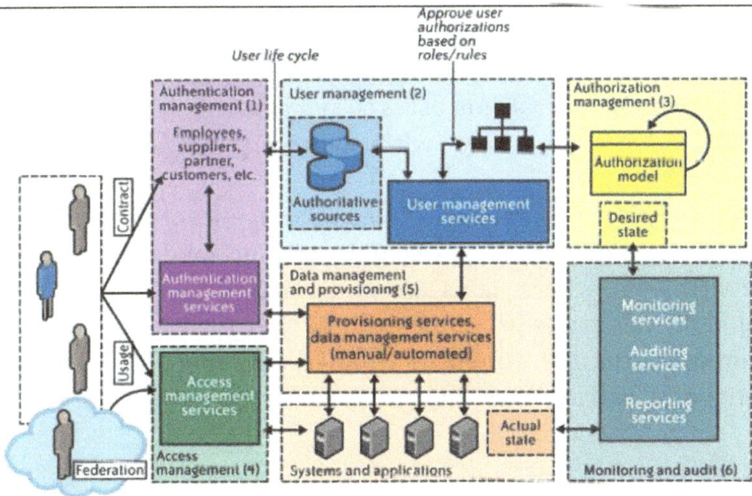

Figure 5.12 Enterprise IAM functional architecture

Standard enterprise IAM architecture encompasses several layers of technology, services, and processes.

- At the core of the deployment architecture is a **directory service** (such as LDAP or Active Directory) that acts as a repository for the identity, credential, and user attributes of the organization's user pool.
- The directory interacts with IAM technology components such as authentication, user management, provisioning, and federation services that support the standard IAM practice and processes within the organization.
- It is not uncommon for organizations to use several directories that were deployed for environment-specific reasons or that were integrated into the environment by way of business mergers and acquisitions.

The IAM processes to support the business can be broadly categorized as follows:

- **User management-** Activities for the effective governance and management of identity life cycles.
- **Authentication management-** Activities for the effective governance and management of the process for determining that an entity is who or what it claims to be.
- **Authorization management-** Activities for the effective governance and management of the process for determining entitlement rights that decide what resources an entity is permitted to access in accordance with the organization's policies.
- **Access management-** Enforcement of policies for access control in response to a request from an entity (user, services) wanting to access an IT resource within the organization
- **Data management and provisioning -** Propagation of identity and data for authorization to IT resources via automated or manual processes.
- **Monitoring and auditing-** Monitoring, auditing, and reporting compliance by users regarding access to resources within the organization based on the defined policies.

IAM processes support the following operational activities:

Provisioning

- This is the process of on-boarding users to systems and applications.
- These processes provide users with necessary access to data and technology resources.
- The term typically is used in reference to enterprise-level resource management.
- Provisioning can be thought of as a combination of the duties of the human resources and IT departments, where users are given access to data repositories or systems, applications, and databases based on a unique user identity.
- Deprovisioning works in the opposite manner, resulting in the deletion or deactivation of an identity or of privileges assigned to the user identity.

Credential and attribute management

- These processes are designed to manage the life cycle of credentials and user attributes create issue, manage, and revoke to minimize the business risk associated with identity impersonation and inappropriate account use.
- Credentials are usually bound to an individual and are verified during the authentication process.
- The processes include provisioning of attributes, static (e.g., standard text password) and dynamic (e.g., one-time password) credentials that comply with a password standard (e.g., passwords resistant to dictionary attacks), handling password expiration, and encryption management of credentials during transit and at rest, and access policies of user attributes (privacy and handling of attributes for various regulatory reasons).

Entitlement management

- Entitlements are also referred to as authorization policies.
- The processes in this domain address the provisioning and deprovisioning of privileges needed for the user to access resources including systems, applications, and databases.
- Proper entitlement management ensures that users are assigned only the required privileges (least privileges) that match with their job functions.
- Entitlement management can be used to strengthen the security of web services, web applications, legacy applications, documents and files, and physical security systems.

Compliance management

- This process implies that access rights and privileges are monitored and tracked to ensure the security of an enterprise's resources.
- The process also helps auditors verify compliance to various internal access control policies, and standards that include practices such as segregation of duties, access monitoring, periodic auditing, and reporting.
- An example is a user certification process that allows application owners to certify that only authorized users have the privileges necessary to access business-sensitive information.

Identity federation management

- Federation is the process of managing the trust relationships established beyond the internal network boundaries or administrative domain boundaries among distinct organizations.
- A federation is an association of organizations that come together to exchange information about their users and resources to enable collaborations and transactions (e.g., sharing user information with the organizations' benefits systems managed by a third-party provider).
- Federation of identities to service providers will support SSO to cloud services.

Centralization of authentication (authN) and authorization (authZ)

- A central authentication and authorization infrastructure alleviates the need for application developers to build custom authentication and authorization features into their applications.
- Furthermore, it promotes a loose coupling architecture where applications become agnostic to the authentication methods and policies.
- This approach is also called an **"externalization of authN and authZ"** from applications.

Figure 5.13 illustrates the identity life cycle management phases.

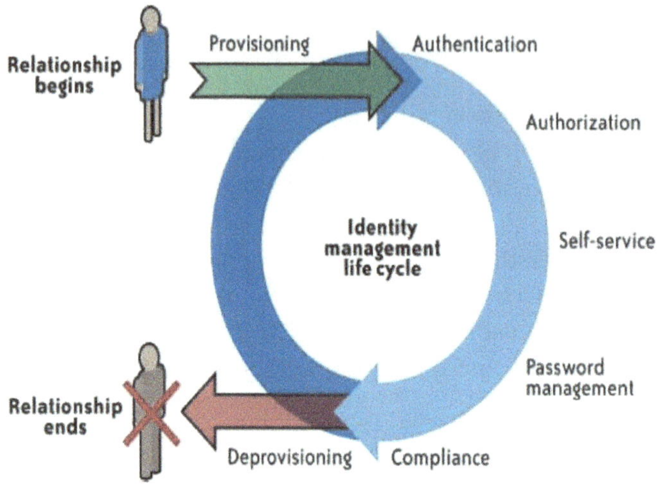

Figure 5.13 Identity life cycle

5.9 IAM practices in the cloud

In the current state of IAM technology, standards support by CSPs (SaaS, PaaS, and IaaS) is not consistent across providers. Although large providers such as Google, Microsoft, and Salesforce.com seem to demonstrate basic IAM capabilities, our assessment is that they still fall short of enterprise IAM requirements for managing regulatory, privacy, and data protection requirements. Table 5.3 illustrates the current maturity model, based on the authors' assessment, generalized across SPI service delivery models.

Level	SaaS	PaaS	IaaS
User management, new users	Capable	Immature	Aware
User management, user modifications	Capable	Immature	Immature
Authentication Management	Capable	Aware	Capable
Authorization management	Aware	Immature	Immature

Table 5.3 comparison of SPI maturity models in the context of IAM

The maturity model takes into account the dynamic nature of IAM users, systems, and applications in the cloud and addresses the four key components of the IAM automation process:

- User Management, New Users
- User Management, User Modifications
- Authentication Management
- Authorization Management

Figure 5.14 defines the maturity levels as they relate to the four key components.

Level	Immature	Aware	Capable	Mature	Industry-leading
User Management, New Users	Manual, ad hoc, with no formal process	Manual, ad hoc, following established processes	Automated where appropriate Disparate processes	Automated using more than one process	Automated using a single provisioning process
User Management, User Modifications	Manual, ad hoc, per application	Manual, ad hoc, per application group	Manual or automated per application group	Automated per class of application and resource	Automated across the application space
Authentication Management	Manual, ad hoc No common security policy	Addressed per application No common authorization mechanism	Common authentication mechanism No common authentication module	Common authentication module Minimal credentials Common security policy	Common authentication mechanism as a component service to applications Common security policy
Authorization Management	Manual, ad hoc No rule-or role-based authorization	Addressed per application No common authorization mechanism	Common service No common module	Common module Application-specific attributes disparately maintained	Common mechanism Centrally managed attributes Support role Rule-based

Figure 5.14 Comparison of maturity levels for IAM components

By matching the model's descriptions of various maturity levels with the cloud services delivery model's (SaaS, PaaS, IaaS) current state of IAM, a clear picture emerges of IAM maturity across the four IAM components. If, for example, the service delivery model (SPI) is "immature" in one area but "capable" or "aware" in all others, the IAM maturity model can help focus attention on the area most in need of attention.

The principles and purported benefits of established enterprise IAM practices and processes are applicable to cloud services, they need to be adjusted to the cloud environment. Broadly speaking, user management functions in the cloud can be categorized as follows:

- Cloud identity administration
- Federation or SSO
- Authorization management
- Compliance management

5.9.1 Cloud identity administration

Cloud identity administrative functions should focus on life cycle management of user identities in the cloud

- provisioning
- Deprovisioning
- identity federation
- SSO
- password or credentials management
- Profile management, and administrative management.

Organizations that are **not capable of supporting federation** should explore cloud-based identity management services.

- This new breed of services usually synchronizes an organization's internal directories with its directory (usually multitenant) and acts as a proxy IdP for the organization.
- By federating identities using either an internal Internet-facing IdP or a cloud identity management service provider, organizations can avoid duplicating identities and attributes and storing them with the CSP.

5.9.2 Federated identity

Organizations planning to implement identity federation that enables SSO for users can take one of the following two paths (architectures):

- Implement an enterprise IdP within an organization perimeter.
- Integrate with a trusted cloud-based identity management service provider.

Both architectures have pros and cons.

5.9.2.1 Enterprise Identity provider

In this architecture, cloud services will delegate authentication to an organization's IdP.

- In this delegated authentication architecture, the organization federates identities within a trusted circle of CSP domains.
- A circle of trust can be created with all the domains that are authorized to delegate authentication to the IdP.

Figure 5.15 illustrates the IDP deployment architecture.

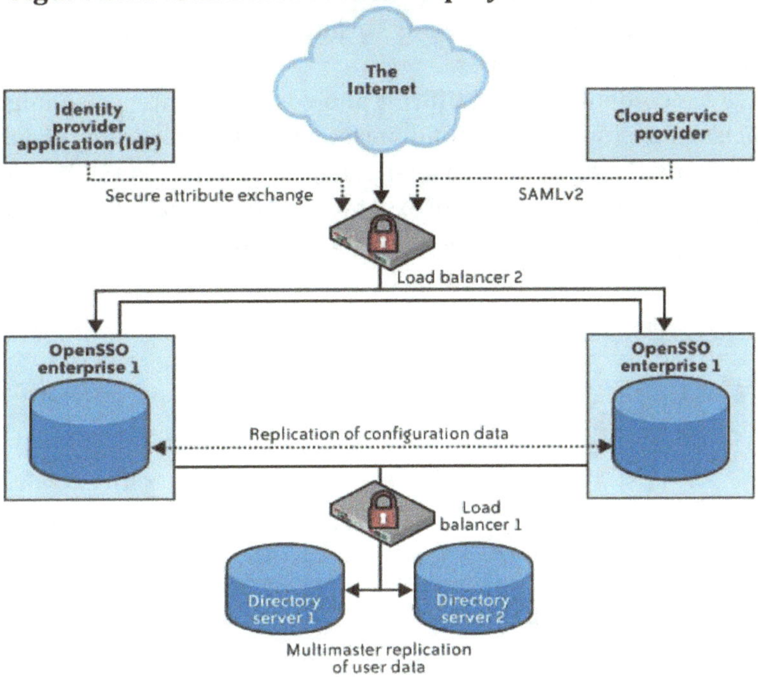

Figure 5.15 Identity provider deployment architecture

Here are the specific pros and cons of this approach:

Pros

Organizations can leverage the existing investment in their IAM infrastructure and extend the practices to the cloud. For example, organizations that have implemented SSO for applications within their data center exhibit the following benefits:

- They are consistent with internal policies, processes, and access management frameworks.
- They have direct oversight of the service-level agreement (SLA) and security of the IdP.
- They have an incremental investment in enhancing the existing identity architecture to support federation.

Cons

By not changing the infrastructure to support federation, new inefficiencies can result due to the addition of life cycle management for non-employees such as customers.

5.9.2.2 Identity management as a service

In this architecture, cloud services can delegate authentication to an identity management-asa- service (IDaaS) provider.

- In this model, organizations outsource the federated identity management technology and user management processes to a third-party service provider.

- the organization might benefit from an outsourced multiprotocol federation gateway (identity federation service) if it has to interface with many different partners and cloud service federation schemes.
- In cases where credentialing is difficult and costly, an enterprise might also outsource credential issuance (and background investigations) to a service provider

This is a SaaS model for identity management, where the SaaS IdP stores identities in a "trusted identity store" and acts as a proxy for the organization's users accessing cloud services, as illustrated in Figure 5.16. The identity store in the cloud is kept in sync with the corporate directory through a provider proprietary scheme

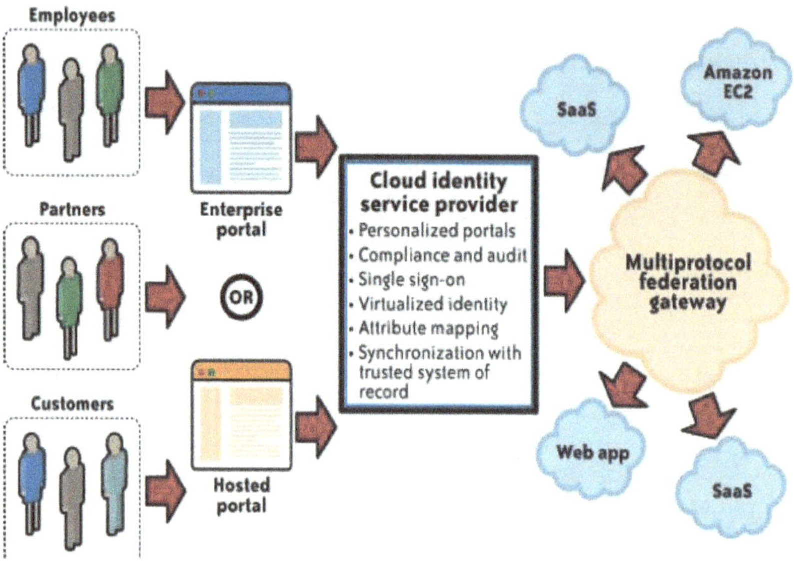

Figure 5.16 Identity management as a service (IDaaS)

Here are the specific pros and cons of this approach:

Pros

Delegating certain authentication use cases to the cloud identity management service hides the complexity of integrating with various CSPs supporting different federation standards.

- Case in point: Salesforce.com and Google support delegated authentication using SAML. However, as of this writing, they support two different versions of SAML: Google Apps supports only SAML 2.0, and Salesforce.com supports only SAML 1.1. Cloudbased identity management services that support both SAML standards (multiprotocol federation gateways) can hide this integration complexity from organizations adopting cloud services.

There is little need for architectural changes to support this model.

Once identity synchronization between the organization directory or trusted system of record and the identity service directory in the cloud is set up, users can sign on to cloud services using corporate identity, credentials (both static and dynamic), and authentication policies.

Cons

Rely on a third party for an identity management service; you may have less visibility into the service, including implementation and architecture details.

- Hence, the availability and authentication performance of cloud applications hinges on the identity management service provider's SLA, performance management, and availability. It is important to understand the provider's service level, architecture, service redundancy, and performance guarantees of the identity management service provider.

Another drawback to this approach is that it may not be able to generate custom reports to meet internal compliance requirements.

In addition, identity attribute management can also become complex when identity attributes are not properly defined and associated with identities.

New governance processes may be required to authorize various operations (add/modify/remove attributes) to govern user attributes that move outside the organization's trust boundary. Identity attributes will change through the life cycle of the identity itself and may get out of sync.

5.9.3 Cloud authorization management

Medium-size and large organizations usually have specific requirements for authorization features for their cloud users. Most cloud services support at least dual roles (privileges): administrator and end user. It is a normal practice among CSPs to provision the administrator role with administrative privileges. These privileges allow administrators to provision and deprovision identities, basic attribute profiles, and, in some cases, to set access control policies such as password strength and trusted networks from which connections are accepted.

5.10 SaaS, PaaS, IaaS availability in the cloud: Factors Impacting availability

The cloud service resiliency and availability depend on a few factors, including the CSP's data center architecture (load balancers, networks, systems), application architecture, hosting location redundancy, diversity of Internet service providers (ISPs), and data storage architecture. Following is a list of the major factors:

- SaaS and PaaS application architecture and redundancy.
- Cloud service data center architecture, and network and systems architecture, including geographically diverse and fault-tolerance architecture.
- Reliability and redundancy of Internet connectivity used by the customer and the CSP.
- Customer's ability to respond quickly and fall back on internal applications and other processes, including manual procedures.
- Customer's visibility of the fault. In some downtime events, if the impact affects a small subset of users, it may be difficult to get a full picture of the impact and can make it harder to troubleshoot the situation.
- Reliability of hardware and software components used in delivering the cloud service.
- Efficacy of the security and network infrastructure to withstand a distributed denial of service (DDoS) attack on the cloud service.

- Efficacy of security controls and processes that reduce human error and protect infrastructure from malicious internal and external threats, e.g., privileged users abusing privileges.

5.10.1 SaaS availability management

SaaS service providers are responsible for business continuity, application, and infrastructure security management processes. This means the tasks your IT organization once handled will now be handled by the CSP.

Challenges of governance of SaaS services as they try to map internal service-level categories to a CSP.

- For example, if a marketing application is considered critical and has a high service-level requirement, how can the IT or business unit meet the internal marketing department's availability expectation based on the SaaS provider's SLA? In some cases, SaaS vendors may not offer SLAs and may simply address service terms via terms and conditions.
- Example: CRM SaaS provider, NetSuite, offers the following SLA clauses:
    - Uptime Goal—NetSuite commits to provide 99.5% uptime with respect to the NetSuite application, excluding regularly scheduled maintenance times.
    - Scheduled and Unscheduled Maintenance Regularly scheduled maintenance time does not count as downtime. Maintenance time is regularly scheduled if it is communicated at least two full business days in advance of the maintenance time. Regularly scheduled maintenance time typically is communicated at least a week in advance, scheduled to occur at night on the weekend, and takes less than 10–15 hours each quarter.
    - NetSuite hereby provides notice that every Saturday night 10:00pm–10:20pm Pacific Time is reserved for routine scheduled maintenance for use as needed.

There is no such thing as standard SLA among cloud service providers. Uptime guarantee, service credits, and service exclusions clauses will vary from provider to provider.

Customer Responsibility

- Customers should understand the SLA and communication methods (e.g., email, RSS feed, website URL with outage information) to stay informed on service outages.
    - When possible, customers should use automated tools such as Nagios or Siteuptime.com to verify the availability of the SaaS service.
- Customers of a SaaS service have a limited number of options to support availability management.
    - Hence, customers should seek to understand the availability management factors, including the SLA of the service, and clarify with the CSP any gaps in SLA exclusions and service credits when disruptions occur.
- The efficacy of SaaS SLAs was analyzed in the context of software vendors moving to a SaaS delivery model.

  - The paper concluded that certain elements are necessary to make the SLA an effective document, and states that: Communication and clear expectations are required from both the service provider and their customers to identify what is important and realistic with respect to standards and expectations.
- Customers of cloud services should note that a multitenant service delivery model is usually designed with a "one size fits all" operating principle
  - This means CSPs typically offer a standard SLA for all customers. Thus, CSPs may not be amenable to providing custom SLAs if the standard SLA does not meet your service-level requirements.
- Since most SaaS providers use virtualization technologies to deliver a multitenant service, customers should also understand how resource democratization occurs within the CSP to best predict the likelihood of system availability and performance during business fluctuations.
  - If the resources (network, CPU, memory, storage) are not allocated in a fair manner across the tenants to perform the workload, it is conceivable that a highly demanding tenant may starve other tenants, which can result in lower service levels or poor user experience.

SaaS Health Monitoring

The following options are available to customers to stay informed on the health of their service:

- Service health dashboard published by the CSP. Usually SaaS providers, such as Salesforce.com, publish the current state of the service, current outages that may impact customers, and upcoming scheduled maintenance services on their website.
- The Cloud Computing Incidents Database (CCID).
- Customer mailing list that notifies customers of occurring and recently occurred outages.
- Internal or third-party-based service monitoring tools that periodically check SaaS provider health and alert customers when service becomes unavailable (e.g., Nagios monitoring tool).
- RSS feed hosted at the SaaS service provider.

5.10.2 PaaS availability management

- In a typical PaaS service, customers (developers) build and deploy PaaS applications on top of the CSP-supplied PaaS platform.
  - The PaaS platform is typically built on a CSP owned and managed network, servers, operating systems, storage infrastructure, and application components (web services).
- Given that the customer PaaS applications are assembled with CSP-supplied application components and third-party web services components (mash-up applications), availability management of the PaaS application can be complicated.
  - for example, a social network application on the Google App Engine that depends on a Facebook application for a contact management service.

- The customer is responsible for managing the availability of the customer developed application and third-party services, and the PaaS CSP is responsible for the PaaS platform and any other services supplied by the CSP.
    - For example, Force.com is responsible for the management of the App Exchange platform, and customers are responsible for managing the applications developed and deployed on that platform.
- By design, PaaS applications may rely on other third-party web services components that are not part of the PaaS service offerings
    - hence, understanding the dependency of your application on third-party services, including services supplied by the PaaS vendor, is essential.
- PaaS providers may also offer a set of web services, including a message queue service, identity and authentication service, and database service, and your application may depend on the availability of those service components (an example is Google's BigTable).
    - Hence, your PaaS application availability depends on the robustness of your application, the PaaS platform on which the application is built, and third-party web services components.
- Customers are encouraged to read and understand the PaaS platform service levels (if available), including quota triggers that may limit resource availability for their application. Two kinds of quotas: a billable quota or a fixed quota.
    - **Billable quotas** are resource maximums set by you, the application's administrator, to prevent the cost of the application from exceeding your budget. Every application gets an amount of each billable quota for free. You can increase billable quotas for your application by enabling billing, setting a daily budget, and then allocating the budget to the quotas. You will be charged only for the resources your app actually uses, and only for the amount of resources used above the free quota thresholds.

    Fixed quotas are resource maximums set by the App Engine to ensure the integrity of the system. These resources describe the boundaries of the architecture, and all applications are expected to run within the same limits. They ensure that another app that is consuming too many resources will not affect the performance of your app.

Customer Responsibility

The PaaS application customer should carefully analyze the dependencies of the application on the third-party web services (components) and outline a holistic management strategy to manage and monitor all the dependencies.

The following considerations are for PaaS customers:

PaaS platform service levels - Customers should carefully review the terms and conditions of the CSP's SLAs and understand the availability constraints.

Third-party web services provider service levels- When your PaaS application depends on a third-party service, it is critical to understand the SLA of that service.

PaaS health Monitoring

- PaaS applications are always web-based applications hosted on the PaaS CSP platform.
    - Hence, most of the techniques and processes used for monitoring a SaaS application also apply to PaaS applications.
- Given the composition of PaaS applications, customers should monitor their application, as well as the third-party web component services.
    - Configuring your management tools to monitor the health of web services will require the knowledge of the web services protocol (HTTP, HTTPS) and the required protocol parameters (e.g., URI) to verify the availability of the service.

The following options are available to customers to monitor the health of their service:

- Service health dashboard published by the CSP (e.g., http://status.zoho.com)
- CCID (this database is generally community-supported, and may not reflect all CSPs and all incidents that have occurred)
- CSP customer mailing list that notifies customers of occurring and recently occurred outages
- RSS feed for RSS readers with availability and outage information
- Internal or third-party-based service monitoring tools that periodically check your PaaS application, as well as third-party web services that monitor your application (e.g., Nagios monitoring tool)

5.10.3 IaaS Availability Management

- Availability considerations for the IaaS delivery model should include both a computing and storage (persistent and ephemeral) infrastructure in the cloud.
    - IaaS providers may also offer other services such as account management, a message queue service, an identity and authentication service, a database service, a billing service, and monitoring services.
- Hence, availability management should take into consideration all the services that you depend on for your IT and business needs.
    - Customers are responsible for all aspects of availability management since they are responsible for provisioning and managing the life cycle of virtual servers.

Managing your IaaS virtual infrastructure in the cloud depends on five factors:

- Availability of a CSP network, host, storage, and support application infrastructure. This factor depends on the following:
    - ➔ CSP data center architecture, including a geographically diverse and fault-tolerance architecture.
    - ➔ Reliability, diversity, and redundancy of Internet connectivity used by the customer and the CSP.
    - ➔ Reliability and redundancy architecture of the hardware and software components used for delivering compute and storage services.
    - ➔ Availability management process and procedures, including business continuity processes established by the CSP.

- ➔ Web console or API service availability. The web console and API are required to manage the life cycle of the virtual servers. When those services become unavailable, customers are unable to provision, start, stop, and deprovision virtual servers.
- ➔ SLA. Because this factor varies across CSPs, the SLA should be reviewed and reconciled, including exclusion clauses.

- Availability of your virtual servers and the attached storage (persistent and ephemeral) for compute services.
- Availability of virtual storage that your users and virtual server depend on for storage service.
  - ➔ This includes both synchronous and asynchronous storage access use cases. **Synchronous storage access** use cases demand low data access latency and continuous availability, whereas asynchronous use cases are more tolerant to latency and availability. Examples for synchronous storage use cases include database transactions, video streaming, and user authentication. Inconsistency or disruptions to storage in synchronous storage has a higher impact on overall server and application availability. A common example of an **asynchronous use case** is a cloud-based storage service for backing up your computer over the Internet.
- Availability of your network connectivity to the Internet or virtual network connectivity to IaaS services. In some cases, this can involve virtual private network (VPN) connectivity between your internal private data center and the public IaaS cloud (e.g., hybrid clouds).
- Availability of network services, including a DNS, routing services, and authentication services required to connect to the IaaS service.

IaaS health Monitoring

The following options are available to IaaS customers for managing the health of their service:

- Service health dashboard published by the CSP.
- CCID (this database is generally community-supported, and may not reflect all CSPs and all incidents that have occurred.
- CSP customer mailing list that notifies customers of occurring and recently occurred outages.
- Internal or third-party-based service monitoring tools (e.g., Nagios) that periodically check the health of your IaaS virtual server. For example, Amazon Web Services (AWS) is offering a cloud monitoring service called Cloud Watch. This web service provides monitoring for AWS cloud resources, including Amazon's Elastic Compute Cloud (EC2). It also provides customers with visibility into resource utilization, operational performance, and overall demand patterns, including metrics such as CPU utilization, disk reads and writes, and network traffic.
- Web console or API that publishes the current health status of your virtual servers and network.

5.11 Key privacy issues in the cloud

Privacy advocates have raised many concerns about cloud computing. These concerns typically mix security and privacy. Here are some additional considerations to be aware of:

Access

- Data subjects have a right to know what personal information is held and, in some cases, can make a request to stop processing it.
- This is especially important with regard to marketing activities; in some jurisdictions, marketing activities are subject to additional regulations and are almost always addressed in the end user privacy policy for applicable organizations.
- In the cloud, the main concern is the organization's ability to provide the individual with access to all personal information, and to comply with stated requests.
- If a data subject exercises this right to ask the organization to delete his data, will it be possible to ensure that all of his information has been deleted in the cloud?

Compliance

- What are the privacy compliance requirements in the cloud?
- What are the applicable laws, regulations, standards, and contractual commitments that govern this information, and who is responsible for maintaining the compliance?
- How are existing privacy compliance requirements impacted by the move to the cloud?
- Clouds can cross multiple jurisdictions; for example, data may be stored in multiple countries, or in multiple states within the United States.
- What is the relevant jurisdiction that governs an entity's data in the cloud and how is it determined? Storage where is the data in the cloud stored? Was it transferred to another data center in another country? Is it commingled with information from other organizations that use the same CSP?
- Privacy laws in various countries place limitations on the ability of organizations to transfer some types of personal information to other countries.
- When the data is stored in the cloud, such a transfer may occur without the knowledge of the organization, resulting in a potential violation of the local law.

Retention

- How long is personal information (that is transferred to the cloud) retained?
- Which retention policy governs the data?
- Does the organization own the data, or the CSP?
- Who enforces the retention policy in the cloud, and how are exceptions to this policy (such as litigation holds) managed?
- Destruction How does the cloud provider destroy PII at the end of the retention period?
- How do organizations ensure that their PII is destroyed by the CSP at the right point and is not available to other cloud users?
- How do they know that the CSP didn't retain additional copies?
- Cloud storage providers usually replicate the data across multiple systems and sites—increased availability is one of the benefits they provide.
- This benefit turns into a challenge when the organization tries to destroy the data—can you truly destroy information once it is in the cloud?

- Did the CSP really destroy the data, or just make it inaccessible to the organization?
- Is the CSP keeping the information longer than necessary so that it can mine the data for its own use?

Audit and monitoring

- How can organizations monitor their CSP and provide assurance to relevant stakeholders that privacy requirements are met when their PII is in the cloud?

Privacy breaches

- How do you know that a breach has occurred, how do you ensure that the CSP notifies you when a breach occurs.
- Who is responsible for managing the breach notification process (and costs associated with the process)?
- If contracts include liability for breaches resulting from negligence of the CSP, how is the contract enforced and how is it determined who is at fault?